Total Energy Independence
for the United States

Total Energy Independence for the United States

A Twelve-Year Plan

ROBERT M. WICAL

iUniverse, Inc.
New York Lincoln Shanghai

Total Energy Independence for the United States
A Twelve-Year Plan

iUniverse books may be ordered through booksellers or by contacting:

iUniverse
2021 Pine Lake Road, Suite 100
Lincoln, NE 68512
www.iuniverse.com
1-800-Authors (1-800-288-4677)

Because of the dynamic nature of the Internet, any Web addresses or links contained in this book may have changed since publication and may no longer be valid.

ISBN: 978-0-595-43519-7 (pbk)
ISBN: 978-0-595-91128-8 (cloth)
ISBN: 978-0-595-87845-1 (ebk)

Printed in the United States of America

<u>Contents</u>

Preface

Most people make critical decisions regarding how to spend their disposable income throughout their adult lives. Nations, too, make critical decisions about what to do with their disposable assets. George W. Bush made an arbitrary national asset allocation decision when he started the latest war in Iraq. It is projected that from its beginning on March 20, 2003, through the year 2007, the United States will spend approximately $500 billion on the war. For what else might this $500 billion have been spent? What if this money had been spent on making the United States both oil independent and well on its way to having a hydrogen fuel infrastructure?

Critical Forks in Roads to the Future

What we do with our world now will dramatically affect the destiny of our descendants. It is well within our power to destroy our civilization, our species, and even our planet. If we capitulate to superstition or greed or politics or stupidity, we can plunge our planet into primitiveness deeper than the Stone Age, or, even worse, we can totally destroy our atmosphere. Two imperatives for the survival of Earth as we know it are global population control to the extent that we do not exceed the Earth's ability to support humanity and global implementation of a non-fossil energy infrastructure that is conducive to restoring and preserving our atmosphere.

With respect to energy independence, the United States of America is at a fork in the road to its energy future: we can either risk annihilation by continuing to rely on oil supplied by countries with which we have an adversarial relationship, or the United States can somehow become self-reliant for all of its oil needs—in other words, oil independent. For the sake of national security, we must achieve oil independence very soon.

Fortunately, oil independence for the United States is achievable and it can be done in about six or seven years, if given the requisite leadership and funding. The hydrogen fuel infrastructure is achievable in an additional six years, given the same caveats. Hence, the original title of this book: *Total Energy Independence for the United States in Twelve Years*. The real significance of an energy-independent United States is that the world's energy requirements are expected to double by 2050, and the likelihood of energy wars will probably double by 2050.[1] If the United States wants to avoid energy wars, achieve oil independence, and maximize its national security, then a hydrogen fuel infrastructure is the way to go.

The notes for each chapter are located at the end of this book. These notes contain Internet addresses that can substantiate and furnish additional information on the technologies cited. This information will help validate the data presented here, and it can also enhance understanding of subsequent information. I chose online sources for this book because Internet access is available to nearly everyone at a very low cost through the use of home computers, school computers, and computers at public libraries.

Introduction

President George W. Bush is referred to here as "Bush," and Vice President Richard Cheney is referred to here as "Cheney." The administration of George W. Bush is referred to here as "Bush & Co."

Early in 2001, the first year of Bush & Co.'s existence, Cheney met with the management of the United States' primary energy-producing companies to develop a "national energy strategy." No one outside of Bush & Co. knows what the national energy strategy is, because Bush & Co. refused to tell the rest of the nation what Cheney and the companies' management decided. Note, however, that the price of oil on the world market began to increase rapidly within a few weeks after this meeting. In January 2001, oil was selling for about $25.35 per barrel (in 2006 dollars). In August 2006, the price of oil had risen to over $78 per barrel—an increase of about $53 per barrel in five years.[1]

Thirty years before September 11, 2001, it was glaringly apparent to anyone who was paying attention that the United States was becoming more dependent on unfriendly nations for its oil supply. During the 1970s, many of these same adversarial oil-producing nations conducted an embargo that caused oil prices throughout the world to reach record highs. In response to these high prices, our national leaders enjoined industries and individuals to find ways to cut oil consumption and to use alternative energy sources. As a result, Congress passed the 55-mile-per-hour national speed limit law and established the Corporate Average Fuel Economy (CAFE) standards program. The CAFE standards program is supposed to be a mechanism for Congress, not the president, to regulate the efficiency of cars and light trucks sold in the United States.[2]

The CAFE standard for cars has been 27.5 miles per gallon since 1990. One reason the standard has not changed is that automobile manufacturers usually do not want to pay the cost of developing more efficient vehicle drive trains. Another reason the standard has not changed is that automotive fuel producers usually do not favor the reduction of fuel consumption, because of the adverse impact on their profits.

As of September 2006, the cost of the Iraq War was estimated to be about $320 billion, and Congress reportedly will appropriate another $170 billion in 2007. The current estimate is that the total cost will be $500 to $800 billion by the time all coalition forces are withdrawn. The Iraq War is the second of two energy wars that have been fought in Iraq; the Gulf War (1990 to 1991) was the first energy war, fought primarily in Iraq and Kuwait.

There was and still is a better alternative to starting a war in Iraq and subsequently killing and maiming hundreds of thousands of people and destroying hundreds of billions of dollars worth of Iraqi infrastructure just to keep a foot in the door of the "Middle East House of Oil." Rather than implement a foreign policy of democracy by invasion, why not use the United States' cash resources to start creating oil independence and a hydrogen fuel infrastructure?

According to my estimates of costs associated with implementing the specific plan described in this book, the initial cost of the hydrogen fuel infrastructure that needs to be created in the United States will likely be in the neighborhood of $1.25 to $1.5 trillion. Ninety percent of this cost will be self-funding if the project is implemented in a manner similar to the one described in this book.

If the United States becomes energy independent, consideration must be given to the following:

• With the United States no longer being a net oil importer, there would be a worldwide impact.

- The security of the nation would be greatly enhanced if the United States became a net oil producer.

- American oil companies would have to accommodate to many operational changes after transitioning to a hydrogen fuel infrastructure.

- A large number of good jobs would be created in the course of constructing a hydrogen fuel infrastructure.

The number of good jobs that would be created could even allow the U.S. economy to start recovering from the "Great Bush & Co. Job Giveaway."

Apparently, Bush & Co. did not give much thought to the effect that a 25 percent reduction in the world's demand for oil would have on Middle Eastern countries, whose entire economy is based on oil revenues. It is also quite likely that Bush & Co. gave little thought to how to produce the vast amounts of hydrogen that will be needed in a United States with a hydrogen fuel infrastructure, and what would be the effects of the absence of millions of tons of carbon dioxide (a greenhouse gas) and carbon monoxide in the Earth's atmosphere. Might the rate of increase in Earth's temperature begin to decelerate?

There are people who profess a disbelief in global warming. These people are in denial—as Ross Perot said of the politicians of his time, "They just want to keep the music playing until the dancehall burns." In addition, there are people who think that it is already too late to do anything about global warming. However, most people who acknowledge the fact that the Earth is getting warmer feel that "better late than never" is a better attitude and a better way to approach the problem.

The Bush & Co. strategy for the Iraq War has been "Stay the Course."[3] This strategy implied more casualties and more human suffering. The national energy strategy established by Cheney *et al.* was probably to stay the course with fossil fuels, but raise the prices. Bush & Co. apparently gave little thought to a hydrogen fuel infrastructure option that would solve the problem of the United States' oil addiction and simultaneously enhance its security.

The hydrogen fuel infrastructure option as described in this book is not hindsight. It was a viable option to any president and congress since 1997, though it probably would have taken one or two years longer to realize than the plan presented here. Yet here we are, ten years later, and there is still little or no progress in curing the nation's oil addiction, despite the fact that a hydrogen fuel infrastructure is an even more viable option today than it was ten years ago.

What is the reason no one has proposed an energy independence plan using the proven energy alternatives available to the United States today? Is it simply a matter of no one taking the time to see the progress of the technologies involved? The answer to these questions may lie in my assertion that few people either in Bush & Co. or in Congress care more about the welfare of the United States than they do about their own welfare and the welfare of Big Business. After all, Big Business's lobbyists are not paying congresspersons to care more about the welfare of the United States! Some might think that this last statement is mere cynicism, but it is not—it is purely the truth.

In the course of conducting research on the current state of hydrogen-related technologies, what was *not* apparent from the literature was any semblance of a *focused, funded plan* for the United States to achieve total energy independence. According to Merriam-Webster's dictionary: "a plan is a method devised for making or doing something or achieving an end." The dictionary goes on to say that the word "plan" always implies mental formulation and sometimes implies graphic representation. There are energy-related documents in the public domain that people refer to as plans, but in reality they are mostly drafts or proposals for what may some-day become plans. In the aggregate, these are basic proposals to develop everything even slightly related to a type of alternative energy, and they have not identified a way to pay for the research, development, and implementation of these plans. There is rarely any mention made of the synergy that could accrue from an aggregation of several of the hydrogen fuel technologies, nor is there any suggestion of a plan explaining how to implement a hydrogen fuel infrastructure. Allocation of funds by a president to perform research and development on a wide variety of alternative energy sources does not constitute a plan for energy independence.

There is one draft of a plan proposing to reduce the United States' reliance on imported oil. This plan uses the free enterprise concept and a wide variety of alternative energy initiatives as an approach to achieving its goals. This draft is progressing toward its goals as fast as corporate greed will allow and perhaps will result in energy independence by the year 2040.[4]

The consensus derived from United States government literature currently on the Internet is that the United States will be using hydrogen as an available vehicle fuel in the year 2030, as part of a national initiative that still requires a detailed definition. Exactly how the Department of Energy intends to implement a hydrogen fuel infrastructure is nebulous at best. There are many stopgap plans to mitigate the United States' oil addiction and minimize greenhouse gas emissions, but the plan I introduce in this book is currently the only known public one that defines and discusses a viable, permanent cure for the United States' energy problems.[5]

The goal of this book is to introduce one of many possible approaches to achieving both oil independence and construction of a hydrogen fuel infrastructure in the United States in a span of twelve years. The approach I present describes a two-phase plan that, if started in the year 2008, could yield both oil independence and a hydrogen fuel infrastructure in the United States by 2021. No, this plan is not science fiction. It will take a herculean effort to implement the Twelve-Years-to-Hydrogen Plan in the time specified, but it would be worth the effort. The United States must adopt some plan for oil independence and a non-fossil fuel infrastructure very soon—we do not have much time to reduce the causes of global warming. We certainly do not have as much time as the global warming skeptics would like to think we have. Nor do we have much time to achieve oil independence before the anti-West Islamists in Iran have nuclear weapons.

Current Helter-Skelter Alternative Energy Initiatives

One of the cleaner alternatives proposed by hydrogen skeptics is clean coal. The fact that "clean coal" is an oxymoron should raise some questions, as should the fact that the sulfur and nitrogen byproducts from burning coal are some of the hazardous components in acid rain. In the mid-to-late 1990s, the acidity of the rain in the Washington DC area was occasionally about the same as the acidity of tomato juice, which has an acidity level—or ph—of 4.1.[5a] Allegedly, this acidity was due to the sulfur compounds released into the atmosphere by coal-burning power plants hundreds of miles west of Washington DC.

In December 2006, in response to Bush's directive to develop a hydrogen economy by drawing upon the best scientific research to address the issue of global climate change, the secretary of energy, Samuel Bodman, announced an agreement with the FutureGen Industrial Alliance to build a prototype of the fossil-fueled power plant of the future. The nearly $1 billion joint government-industry project will produce electricity and hydrogen with zero emissions of carbon dioxide, a greenhouse gas. The *coal-fed* plant will reportedly produce electricity to make hydrogen, and will lock away the carbon dioxide in unproductive oil wells to achieve what it bills as "the world's first zero-emissions fossil fuel plant."[6] Consider also the $360 million natural gas reformer the Department of Energy funded for Chevron Mobile in 2003, as a hydrogen source. A natural gas reformer is a plant designed to extract hydrogen from natural gas and then release the remaining carbon dioxide and other gases into the atmosphere.

Then there is Bush's Hydrogen Fuel Initiative, which calls for replacing fossil-fueled passenger cars with hydrogen-powered cars in the 2020 to

2030 time frame. However, this is mostly unfunded speculation. Bush's Hydrogen Fuel Initiative would require producing 160 to 170 million tons of hydrogen annually, and there is no apparent plan or funding to accomplish this goal.[7] There are government and private sector–funded solar, wind, clean coal, biomass, and nuclear alternative energy initiatives in progress as well. Most of the current recommended remedies for our oil addiction involve conservation. Of course, we still have the fossil fuel advocates, who expound about the oil reserves in this oil field and the oil reserves in that oil field[8]—meanwhile, planet Earth and the United States are becoming more and more in harm's way because of the *world's* addiction to fossil fuels.

With so many alternative energy initiatives in progress simultaneously, it is difficult to determine the true state of development of any single alternative energy initiative. By distributing the funds allocated for the development of alternative energy sources over so many different initiatives, there is not enough money left to concentrate on the development of the most promising alternative energy initiatives. Currently, the only focus in alternative energy development is on all of the possible alternative energy initiatives simultaneously; there is no focus on developing a plan for *oil independence or for a hydrogen fuel infrastructure*. My research indicates that the cost estimates for developing a hydrogen fuel infrastructure in the United States have ranged from $840 billion to $22 trillion, using any one of the six leading alternative energy technologies.[9]

The reason for such cost disparities is that there is no plan that clearly defines and describes what constitutes the primary components of a hydrogen fuel infrastructure and how many of each component will be required. There are many other disparities in the available data on alternative energy initiatives; misinformation is abundant. Therefore, one of the most important tasks in developing a plan for an oil-independent United States—and to subsequently develop a plan for a hydrogen fuel infrastructure—is to determine which technologies are the essential, or *core*, technologies needed to attain both oil independence and a hydrogen fuel infrastructure. After identifying the core technologies, the next step should be to determine the current state of development of each of the core technologies.

It could take many years to review all the Internet data related to hydrogen fuel technology. The technologies referred to here as "core technologies" are the ones that should receive the bulk of the research and development resources. The amount of existing detailed information pertaining to each of the core technologies is extensive, and widely dispersed over a multitude of Web sites.

The primary objective here is to make information and ideas available so more people will be aware of the state of scientific developments that can lead to oil independence and a hydrogen fuel infrastructure for the United States. More descriptively, if an economy were deriving its energy from both nuclear energy and hydrogen energy, then perhaps it would be more correct to call it a "Nuclear + Hydrogen fuel infrastructure." Regardless of what the economy is called, the key technological development that will propel the United States to oil independence and a hydrogen fuel infrastructure is a fourth-generation, liquid-metal-cooled nuclear reactor called an Integral Fast Reactor (IFR), or alternatively, a Secure Transportable Autonomous Reactor (STAR).

To help map the direction in which we are heading, a summary of the Twelve-Years-to-Hydrogen Plan will follow. The plan comprises a set of core technologies. Identifying and discussing the developmental states and capabilities of these core technologies, and how they fit into the Twelve-Years-to-Hydrogen Plan constitutes the essence of this book. The core technologies are as follows:

- *Integral Fast Reactor (IFR) and STAR*
  Fourth-generation nuclear reactors are truly safe, relatively inexpensive, and highly efficient, and they can furnish huge amounts of electrical power. The cost of electrical power is currently the main deterrent to establishing a hydrogen fuel infrastructure in the United States. Fortu-

nately, there is already enough nearly free IFR fuel on hand to meet all of the United States' electrical power and hydrogen needs for the next 500 to 1,000 years. See Chapter 1.

- *In situ Conversion Process (pertains to oil shale fields)*
  There are approximately one trillion barrels of oil locked in oil shale (rock) in the West-Central United States; this conversion process drives the oil and natural gas out of the shale without taking the shale out of the ground.[10]

- *Electrolysis of Water*
  Electrolysis of water splits water into its constituent elements: hydrogen and oxygen. Hydrogen can be liberated from water using electrolysis and then used for fuel.[11]

- *Electrochemical Hydrogen Compressor*
  This hydrogen compressor has no moving parts, yet it can compress hydrogen up to 5,000 to 10,000 pounds per square inch, or more, by simply applying voltages to the device.[12]

- *Hydrogen Distribution Systems*
  These systems transport hydrogen from its point of manufacture to the point-of-sale storage tank.[13]

- *Hydrogen Dispensing Systems*
  These systems dispense hydrogen from the point-of-sale storage tank into the storage tank of the vehicle that will consume the hydrogen.[14]

- *Hydrogen-Powered Automobile and Transit Technology*
  Cars and buses that run on hydrogen and employ devices like the fuel cell to power the vehicle are representative of this technology.[15]

- *Proton Exchange Membrane (PEM) Fuel Cell Technology*
  A PEM fuel cell is a device that converts hydrogen to electrical power,

heat, and water vapor; the electrical power from this device can power vehicles or provide power for other commercial purposes.[16]

- *Parallel-Path Magnetic Technology*
This technology uses magnets to increase the efficiency of electric motors and generators, thereby improving the efficiency of vehicles powered by electric motors; for example, fuel cell–powered vehicles.[17]

The Twelve-Years-to-Hydrogen Plan ("Plan") will be accomplished in two phases:

- Phase One: the oil independence phase

- Phase Two: the hydrogen fuel infrastructure phase.

There is a symbiotic relationship between phase one and phase two of the Plan. The tax revenues from phase one will finance phase two, while the oil from phase one satisfies the nation's oil addiction and enhances national security by making the nation self-reliant for its oil supply. This Plan is currently the most cost effective, expedient, and safest plan publicly available.

One of the objectives of the following chapters is explaining how the core technologies interact to achieve the goals of the Plan. As an example, in phase one, the IFRs will provide electrical power to furnish heat to the in situ conversion process. After heating the oil shale fields sufficiently, the oil and natural gas can be removed from the oil shale fields by drilling conventional oil wells. In phase two of the Plan, IFRs will provide the electrical power required to produce all of the hydrogen needed for the hydrogen fuel infrastructure.

A word of caution: when conducting your own research on hydrogen technologies, be aware of the referenceless misinformation that is rampant in extant writings about using hydrogen as a fuel. In addition, be sure to check the dates of the articles you read and try to corroborate the data in another article.

Chapter One

What Is an Integral Fast Reactor?

You might be wondering how Bush & Co. could have converted the United States from an oil-based economy to a hydrogen fuel infrastructure for about $1.25 to $1.5 trillion, and made the United States oil independent in the process. What are the key transitioning processes that can take the United States from an oil-based economy to a hydrogen fuel infrastructure?

You should be forewarned that at various points throughout this book, the verbiage will sound like a course on the fundamentals of nuclear physics; this is especially true of Chapter 1. The reason for discussing nuclear physics is to attempt to explain the fundamental operating characteristics of the device that will produce the electricity needed to create our hydrogen. This device is a grossly misunderstood type of nuclear reactor known as the **Integral Fast Reactor** (IFR) and one of its components, the **Advanced Liquid-Metal Reactor** (ALMR). These two terms are frequently used interchangeably because the ALMR is the nuclear reactor component of the IFR. Detailed characteristics of the IFR and the ALMR will be discussed later.[1]

Because of the design characteristics of the 103 or so thermal or light-water reactors (LWRs) that have been pressed into service in the United States, and the problems that have occurred with thermal reactors, Three Mile Island, and Chernobyl, nuclear power has acquired a bad name in the minds of most Americans. You might think that nuclear reactors are too dangerous and unstable to be relied upon as safe sources of energy. You might know also that each light-water nuclear reactor producing one billion watts of electricity *annually* produces about 17 to 25 tons of highly radioactive waste that must be safely stored for hundreds of thousands of years.[2] Please read on for an explanation of how IFRs, or "fast" reactors, are inherently safe, very clean, and produce only small amounts of radioactive waste that will be safe in 300 years instead of thousands of years—but first, a bit of IFR history.

Argonne National Laboratories and the EBR II

Research on the IFR began in 1984 at Argonne National Laboratory in Argonne, Illinois. Argonne is a part of the U.S. Department of Energy and is co-operated by the Department of Energy and the University of Chicago. Argonne National Laboratory also has a branch at Idaho Falls, Idaho. At the Idaho branch, physicists from Argonne had built what was known as the Experimental Breeder Reactor II (EBRII). At about the same time, physicists at Argonne had designed the IFR concept, and it was decided that the EBRII would be converted to an IFR.[3] The IFR project was a power reactor development program built around a revolutionary concept for generating nuclear power. Not only was it a new type of reactor, it also had an entirely new nuclear fuel cycle.[10a]

Politics and Ignorance Reign Supreme

President Clinton announced in his 1993 State of the Union address that his administration would be eliminating programs that were no longer needed, such as nuclear power research and development programs.[3a] The integral fast reactor (EBR II) was one of the programs selected to be eliminated. Reportedly, Senator John Kerry and Energy Secretary Hazel O'Leary led the opposition to the IFR, arguing that it would be a threat to the United States' non-proliferation efforts, and that it was merely a continuation of the Clinch River breeder reactor effort that had already been cancelled by Congress. Despite support for the IFR by some in Congress, funding for the IFR was slashed, and it was ultimately cancelled in 1994—only three years before project completion. It would have cost less to complete the project than it did to cancel it! The IFR was cancelled because the people who cancelled it did not understand it.

During his 2004 presidential campaign, Senator John Kerry mentioned nuclear power as one of the much-needed alternative energy sources he would fund at the $30 billion level if elected. Ironically, a project like the IFR project would probably have been the type of nuclear power source project he would have funded. Unfortunately, the original Argonne West IFR reactor has been irreversibly dismantled and disabled.[5]

Collectively, our politicians seem to know almost nothing about the hydrogen-related technologies they will have to fund to take the United States into the future. The U.S. Congress tends to rely totally on the testimony of "experts" for technical information. Members of Congress might be given good information, or a biased expert might give them bad information. They truly have no way of knowing if the information they receive is good or bad. Some education of Congress in the field of nuclear energy and hydrogen fuel technology will be necessary.

The Energy of the Future Is Hydrogen

Skeptics say that once the technical obstacles are removed, hydrogen's huge price tag may still make the technology prohibitive.[6] Realistically, however, we must pay the price for hydrogen; we do not have another viable, pro-life energy choice.

A recent analysis by the Department of Energy projected, with a reasonable degree of accuracy, that a hydrogen supply network adequate for even 40 percent of the light-duty fleet (all cars and light trucks) in the United States could cost more than $500 billion.[6] It is also true that developing a hydrogen fuel infrastructure is a classic chicken-and-egg problem. How do you get millions of Americans to buy hydrogen-powered vehicles before there is a hydrogen fuel infrastructure in place to refuel them? And how do you get energy companies to build that hydrogen fuel infrastructure before there's a potential customer base? Obviously, companies are not willing to invest if they do not think there is going to be a market. Therefore, the government must be totally proactive in and totally financially supportive of hydrogen fuel infrastructure construction.

There is little reason to hope that nuclear and hydrogen technologies will advance to the extent needed, and *when* needed, without heavy government involvement—corporate greed precludes it. However, "hydrogen fuel cells already replace batteries in niche equipment, such as TV cameras and forklifts, and provide power at remote locations like cellular phone towers. They even power the police station in New York's Central Park."[9] If you are not familiar with the principles on which the hydrogen fuel cell operates, please visit the Ballard Power Web site.[7] As these fuel cell applications continue to develop, they just might force advances in technology that will make hydrogen-powered vehicles more feasible. Some skeptics say that even when hydrogen-powered vehicles are available, hydrogen might only make sense for local fleet vehicles that do not require a widespread infrastructure for service and refueling.

The ultimate hydrogen fuel skeptics are "industry experts" who have been purchased by a person or entity with self-serving interests they feel compelled to profess falsely to the world. An example would be someone paid to deceptively testify that a hydrogen fuel infrastructure is unnecessary and a waste of money, because some alternative fossil fuel will be the savior of the planet.[7a] Realistically, the ultimate skeptic does not have much time left to deceive the public about global warming—the polar ice caps are melting.[8]

Bush brought into the national awareness the fact that the United States is addicted to oil. Based on Bush & Co.'s paltry allocation of only $1.7 billion for five years of alternative energy development, it is obvious that Bush intended for the United States to "stay the course" on fossil fuels.[9] Serious, accelerated alternative energy development would require thirty times the $1.7 billion allocated over the five-year period covered by the allocation.

Once the economic advantages of using integral fast reactors have been proven unequivocally, then hydrogen will not be just another item in a large collection of energy alternatives; rather, hydrogen will be the alternative energy of choice. A hydrogen fuel infrastructure will involve investing heavily in a new national infrastructure. Though the price tag will be steep, we can no longer afford oil's environmental, economic, and political drawbacks.

Things Nuclear

Here, I present an overview of some basic nuclear physics concepts to help the reader understand why the IFR is inherently safe and why it is up to 99 times more efficient than the thermal nuclear reactors (LWRs) we now use commercially to provide electrical power. In addition to being safe and efficient, the IFR also overcomes nearly all of the other objections to using nuclear power for commercial purposes.[10]

Bush & Co. did revive some aspects of the IFR project in 2003, and they provided about $180 million to continue IFR-related research—much less than the amount needed if development is to be accelerated. Bush & Co. had allegedly eliminated the funding in their 2003 budget for the thirty-three people who were to perform the IFR-related work. Reportedly, these people were about to be laid off, but when the oversight was publicized, their funding magically reappeared.

Nuclear Power Fuel Cycles Are Important

Traditional nuclear power fuel cycles were derived from technologies developed to meet special military needs: naval propulsion, uranium enrichment, weapons-plutonium production, and plutonium separation.[10b] Historically, radioactive waste disposal has always been approached as "someone else's problem." The IFR concept is directed strictly at meeting the needs of civilian power generation. It has an integrated, weapons-incompatible, proliferation-resistant fuel cycle that is "closed." A closed fuel cycle is one in which the waste fuel does not leave the reactor site. Instead, the spent fuel is reprocessed in reprocessing facilities integrated into the IFR site. With the exception of uranium mining, the IFR fuel cycle encompasses the entire remainder of the nuclear reactor fuel cycle, including fuel production and fabrication, power generation, fuel reprocessing, and radioactive waste management.[11] IFRs will reduce or eliminate nearly all of the significant difficulties that beset today's thermal reactor (LWR) fuel cycle, such as the following:

- Production and buildup of waste plutonium

- Short-term management of waste plutonium

- Disposition and long-term management of plutonium

- Plutonium in national and international commerce

- Other proliferation concerns

- Long-term radioactive waste management

- Environmental effects

- Resource conservation

- Long-term energy supply

- Safety[1]

The nuclear reactor "fuel cycle" refers to the process by which the uranium ore is mined, milled, enriched, processed, used, possibly reprocessed, and then buried as a means of disposal.[12] A description of these steps follows:

- *Uranium Mining*
  Uranium is sampled and mined as other metals are—via open-pit mining or leach mining. Raw uranium ore found in the United States ranges from 0.05 percent to 0.3 percent uranium oxide. Uranium ore is not rare.

- *Uranium Milling*
  Raw ore is milled by grinding and chemically leaching the ore. The resulting natural uranium oxide powder is called "yellowcake." The yellowcake powder is then converted to uranium hexafluoride to prepare it for enrichment.

- *Uranium Enrichment*
  Less than 1 percent of the uranium found in nature is the easily fissionable U-235 isotope. As a result, most reactor designs require enriched fuel. Enrichment involves increasing the percentage of U-235, and is usually done by means of gaseous diffusion or gas centrifuge.

- *Uranium Fabrication*
  The enriched uranium is then converted into uranium dioxide powder, which is pressed and heated into pellet form. These pellets are stacked into tubes, which are then sealed and called fuel rods. Many of these fuel rods are used in each nuclear reactor.

- *Reactor Operation*
  The fuel is used.

- *Fuel Reprocessing*
  Fuel reprocessing methods are discussed below.

- *Radioactive Waste Management*
  The final stage of the nuclear fuel cycle is the management of the still highly radioactive, spent fuel, which constitutes the most problematic component of the nuclear waste stream. After fifty years of producing nuclear power, the question of how to deal with the radioactive waste

material remains fraught with safety concerns and technical problems. One of the most important lines of criticism of the industry is based on the long-term risks and costs associated with dealing with the waste. Later, I will discuss how the IFR eliminates this problem.

Depending on the type of fuel cycle, some of the above steps might not be needed. The three major types of fuel cycles are: thermal without reprocessing ("once-through," or "throw-away"), thermal with reprocessing, and IFR. The IFR fuel cycle will essentially eliminate the future need for uranium mining, uranium milling, uranium enrichment, and nearly all of the waste plutonium disposal requirements.[12]

Nuclear power fuel cycles are so important that the Department of Energy has recently launched the Advanced Fuel Cycle Initiative. The main goals of this initiative are to:

- Reduce the number of geologic nuclear waste repositories required this century to one.

- Reuse valuable parts of spent nuclear fuel to maximize the energy derived from uranium ore.

- Recycle spent nuclear fuel in fast reactors to minimize waste and control weapon-usable inventories.

"The Advanced Fuel Cycle Initiative is a vital part of the Global Nuclear Energy Partnership (GNEP). The GNEP is a comprehensive strategy to enable the expansion of emissions-free nuclear energy worldwide by demonstrating and deploying new technologies to recycle nuclear fuel, minimize waste, and reduce the risk of proliferation of sensitive nuclear technologies. The technologies that are planned to be demonstrated and deployed under GNEP are those most promising technologies that have been developed under the Advanced Fuel Cycle Initiative. The Advanced Fuel Cycle Initiative will now focus on long-term research and development in advanced separation technologies, waste and storage forms, and advanced transmutation fuels. This long-term research will develop technologies that show potential but require further research prior to engineering-scale demonstration and deployment."[13] How the Advanced Fuel Cycle Initiative is going to improve on the Integral Fast Reactor fuel cycle is unclear.

"The Advanced Liquid-Metal Reactor (ALMR) is the reactor component of the IFR; therefore, it is a "fast" reactor. The term "fast reactor" in this context means maintaining the chain reaction by bombarding the fissionable material with high-energy neutrons. The energy spectrum of the high-energy neutrons is said to be "fast." There are over 300 "reactor years" of operating experience with fast reactors."[14,15]

A point of interest: an IFR producing 1,000 megawatts of power produces about 1,700 *pounds* of waste per year. In contrast, "a 1,000-megawatt coal-fired power plant operated 70 percent of the year produces 1,275,000 *tons* of waste—and they still call it "clean coal."[16]

There is a "Slow" Reactor

At the "slow" end of the neutron energy scale is the thermal reactor, or Light Water Reactor (LWR). Light-water is just another name for the type of water we drink. There is such a thing as heavy water, which is also known as deuterium oxide. Deuterium is heavy hydrogen. Almost exclusively, reactors in commercial use today by electrical power plants are of the thermal (LWR) variety—there are about 103 commercial thermal nuclear reactors in use in the United States today and approximately 438 worldwide. These LWRs produce about 16 percent of the world's electrical power. The chain reaction of the thermal reactor relies on "slow" neutrons. In most thermal-spectrum reactors (LWRs), the neutrons are moderated or slowed by light water. This is why the thermal reactor is also called a "Light-Water Reactor" (LWR). It is the problems the power industry has had with these LWRs that have given nuclear power a bad reputation.[1]

Now that we are aware of the fact that there are fast reactors and slow (thermal) reactors, the next question is: what is the most important difference in the capabilities of these two types of reactors?

A Thermal Reactor (LWR) Always Produces Plutonium

The most important difference between the fast reactor and the thermal reactor is that thermal reactors produce copious amounts of plutonium, while an IFR *can* consume the plutonium it produces. In fact, two Argonne-type IFRs could consume the waste plutonium output of five thermal reactors (LWRs) of the same size, and use the energy in the waste plutonium to generate electricity and revenue. In case you are not aware of any of the uses for plutonium, high-quality plutonium is the preferred bomb-making material for a sophisticated nuclear weapons program. It is even possible to make a nuclear explosive with the low-quality plutonium found in thermal electrical power reactors (LWRs). Each LWR produces about 17 to 25 tons of "high-level" radioactive waste each year. IFRs can burn (consume) this radioactive waste and use the resulting energy to make electricity and hydrogen.[1]

IFRs Can Burn Almost Any Nuclear Fuel

In addition to burning (consuming) the various qualities of plutonium other reactors produce as waste, the fast-energy spectrum of IFRs allows them to burn any and all actinides. Actinides are elements with an atomic number of 89 (actinium) and greater. IFRs can burn actinides because in the fast-neutron spectrum all the actinide isotopes have roughly the same fission probability or "fission cross-section." The most important actinide elements are uranium (atomic number 92), plutonium (atomic number 94), and, to a lesser extent, thorium (90). Since there is a growing glut of weapons-grade plutonium piling up from dismantled atomic weapons and ongoing worldwide thermal reactor (LWR) operations, the first IFRs will undoubtedly be fueled with some of the "waste" weapons' plutonium.[15]

In the thermal-neutron spectrum of the LWR, many of the fission products and actinide isotopes absorb neutrons readily without undergoing fission; that is, they have a high capture cross section, and the chain reaction is slowed or stopped (poisoned) if too much of such material is present. It is like mixing water with gasoline to the extent that the gasoline no longer burns. Thus, a thermal reactor cannot be a net burner of transuranic actinides. Transuranic actinides are the elements beyond uranium—that is, their atomic number is 93 or greater: neptunium, plutonium, americium, curium, and others. All transuranic actinides are fabricated elements. They are so radioactive that the naturally created transuranic actinides, also called higher actinides, have long since decayed away in our little bit of the universe.

The main starting fuel for thermal reactors is a mixture of the fissile isotope U-235 (uranium) or Pu-239 (plutonium) along with the fertile isotope U-238 (uranium). The term "fertile" isotope applies when the addition of neutrons to atoms of an isotope changes the non-fissile isotope into a fissile isotope—one that, like U-235, has a very high probability of undergoing fission when exposed to (bombarded by) thermal neutrons. A special note about U-235 is warranted here: the estimated reserve of U-235 is "enough for many decades if only LWRs are used." There is not enough U-235 to last for thousands of years. Both fissile and fertile isotope atoms are fissionable, but the fertile ones require a high-energy neutron to make them split. As you probably know, atoms must be split to produce the infamous chain reactions that produce atomic energy.[16]

Breeder Reactor vs. Integral Fast Reactor

A reactor that is intentionally configured to produce more fissile material than it consumes is called a "breeder" reactor. A fast reactor is designed and operated to be either a net *breeder* of fissile material or a net *burner* of fissile material. A thermal reactor (LWR) is a net burner of nuclear fuel, but—*and this is critically important*—all thermal reactors are inherently prolific breeders of plutonium. A thermal reactor starts out with no plutonium at all, and soon has a lot of it. In the process, though, it burns more fuel (mainly uranium) than it gives back as plutonium, and, therefore, is not called a breeder reactor.

Confusing breeder reactors with IFRs probably started while scientists were investigating IFRs for their potential to breed fissile material. The central fact people seem to miss here is that with IFRs you can choose not to breed plutonium, whereas with thermal reactors you make plutonium whether you want it or not. In other words, it is primarily the thermal reactors (LWRs) being used throughout the world to produce electricity that are the runaway producers of plutonium and that present the greatest risk of proliferating nuclear materials. IFRs could put a stop to the runaway production of plutonium—a most important difference.[1]

Why the Concern about Plutonium?

The weapons-grade plutonium that is removed from dismantled atomic bombs can be degraded to low-quality plutonium by mixing the weapons-grade material into the fuel for today's thermal reactors. The fuel would then consist of a mixture of oxides of uranium and plutonium, called "MOX." This process is now being used in the United States and Russia. The IFRs initially brought on line will first directly consume weapons-grade plutonium for fuel. When the weapons-grade plutonium is gone, the thousands of tons of reactor-grade plutonium that is currently containerized and stored in 125 various locations throughout the nation will be used to fuel IFRs.[1]

Integrated Fuel Reprocessing Facilities

Now that we have covered the term "fast" as it is used in Integral Fast Reactor, we are next going to deal with the term "integral." In the IFR context, "integral" refers to the fact that the spent fuel reprocessing facilities are an integral part of the IFR plant; therefore, the fuel cycle is closed. In other words, spent fuel does not need to be taken off the IFR site to be reprocessed. This is an extremely important characteristic of the IFR, because it means that there is no need to be concerned about additional plutonium shipments to the IFR reactor sites, and there will not be multiple spent fuel shipments from the IFR reactor sites to the spent fuel repositories. As a consequence, commerce in plutonium can be kept to an absolute minimum, and the only plutonium shipments would then be from dismantled weapons and thermal reactors to IFR sites or spent fuel repositories.

An IFR plant will be a "sink" or an end-user for plutonium. Plutonium to be disposed of could be shipped to the IFR, and there, with on-site recycling as needed, it would be processed into fuel and then consumed to produce energy ("burned"). Only trace amounts of plutonium would ever leave an IFR plant. The primary reason so little plutonium would ever leave an IFR plant is that the fuel is reprocessed using a pyroprocess that *cannot* produce plutonium. The actual amount of waste produced by an IFR is so small, it can be stored on the reactor site until the reactor is refurbished or removed.

Another common nuclear fuel reprocessing technique called the PUREX process is used specifically to produce high-grade plutonium—this is definitely what we do not want if we are to discourage plutonium proliferation.[1, 12]

Metallic Fuel Rods Mean Safer Reactors

Although the safety record of commercial reactors of Western design is superb, Three Mile Island aside, it is more desirable to have reactors that rely on inherent safety features and less on engineered ones. One way to create a reactor safety mechanism that does not depend on human or mechanical intervention is to control the form of the material used for fuel. In the ALMR portion of the IFR, the fuel is metallic fuel rods, whereas the thermal reactors (LWRs) use a less dense oxide (nonmetallic) fuel.

Metallic fuel rods, usually an alloy of uranium and plutonium, are an inherent safety feature in the ALMR. This is because metals are good heat conductors, while oxides are poor conductors of heat. The superior thermal conductivity of metal fuel rods means the interiors of the metal fuel rods dissipate their heat faster so they stay much cooler than oxide fuel rods. Another way to explain the effects of the fuel heat retention is that the temperature difference between the reactor coolant and the reactor fuel is less for metallic fuel rods. Therefore, there is much less heat retained in the fuel to raise the temperature of the reactor should the reactor coolant flow be impeded, thereby making the consequences of an accident much less severe. To make a long story short, because the temperature in an ALMR does not change much during the accident scenario, the reactor is a lot more stable. If the heat in the reactor were to increase due to an accident condition, the metallic fuel rods would expand and spread apart at the free end of the rods, like the opening of a flower blossom, and thus cause the chain reaction to slow down automatically because of the decrease in fuel density. The reactor would go subcritical and shut itself down. The ALMR is inherently passively safe.[1]

Cooling by Liquid Metals

Advanced Liquid-Metal Reactors use liquid metals at atmospheric pressure for cooling and heat transfer; a characteristics that make IFRs much safer than LWRs. The first liquid metal used to cool the IFR was sodium. Other metals have been used in ALMRs, including lead, mercury, and lead/bismuth alloy. The LWRs use water under pressure for cooling and heat transfer. If a LWR cooling water pipe should rupture—and if the rupture of the pipe were to be followed by blowdown (escape) of the reactor coolant—the reactor can then overheat and possibly reach the uncontrollable state.

Liquid sodium is not as corrosive as water, but it will burn if exposed to air, and it reacts violently with water. The water/sodium reaction could be explosive if the hydrogen released during the reaction is confined and concentrated. Obviously, the ALMR requires prudent design.[1]

In December 1995, at the Monju ALMR plant in Japan, a sodium leak caused a fire. The plant was damaged and subsequently shut down. Most importantly, though, no one was hurt, no radioactivity was released, and the reactor itself was not damaged. The accident was classified as Category 1 on the international scale of 0 to 7 (0 being the least serious) by a committee of independent specialists. There is a great deal of industrial experience with liquid sodium, and there have been very few problems.[1]

Passive Safety Features Tested

One other inherent safety feature of the IFR lies within the design of the plant. The ALMR core sits in a pool of liquid sodium, and the metal fuel rods have low internal heat content due to their thermal conductivity. These two design characteristics combine in such a way that if there were a loss of reactor control power, the reactor core would be cooled passively by convectively circulated liquid sodium. This design feature of the ALMR has actually been tested. While the IFR was operating at full capacity, the control power for the reactor was turned off, reactor coolant pumps were stopped, the control rods were not moved, and the operators took no remedial action. The core temperature rose slightly, causing the reactor to go subcritical and shut itself down without incident. When a reactor goes sub-critical, it is unable to sustain nuclear fission. Unassisted convective circulation of the Liquid-Metal coolant cooled the core and prevented overheating. Nuclear power cannot get much safer than this.[17]

IFR Fuel for a Thousand Years Already Exists

The level of safety that can be achieved with the IFR is terrific, but the most amazing aspects of the IFR remain to be told. The LWR, with its one-time-through fuel cycle, only uses about 1 percent to 2 percent of the energy in the uranium used to fuel it. In contrast, the IFR will use about 99 percent of its fuel—and its fuel, in most cases, will be the LWR waste plutonium that has been accumulating in caves for over fifty years. The IFR emits no carbon dioxide or other greenhouse gases. In addition, it can be argued that the major environmental problems with nuclear power are the consequences of the uranium mining and milling operations, and radioactive waste disposal. The IFR can use both the surplus plutonium that we have been accumulating in spent fuel repositories for many years, and the uranium that has already been mined and milled and is currently being stored. Because there is so much IFR fuel already in existence, IFRs can eliminate any further need for mining and milling nuclear fuels for up to ten centuries. In other words, we have already stored thousands of tons of waste plutonium from LWRs and dismantled bombs, and we have stored so much milled uranium that we have sufficient fissionable material to meet the energy needs of the United States for at least the next 500 years, and possibly for the next 1,000 years. This would be true even if we have 1,000 IFRs in use.[18] The total amount of waste from an IFR after sixty years of operation will be about 1,700 *pounds*, compared to 17 tons or more *per year* for the LWR.[16]

Greatly Reduced Radioactive Waste

Another advantage of the IFR is that the small amount of radioactive waste it does produce will decompose to a relatively safe level of radioactivity in about 300 years. The relatively large volume of waste plutonium from an LWR will take 100,000 to 300,000 years to decay to a relatively safe level of radioactivity. Again, the IFR eliminates a problem associated with nuclear power. The IFR's characteristics are also consistent with the

United States' policy of not proliferating plutonium. Once the IFRs are in use, useful quantities of weapons-grade plutonium cannot be created in the IFR plant, and no useful quantities of weapons-grade plutonium will ever leave the IFR plant.

Thermal reactors need to have their fissionable fuel replaced (be refueled) about every five years. Refueling a reactor is a very costly process. Because the IFR uses its fuel so completely, the IFR does not have to be refueled. Fueling of the IFR will occur only once—at the beginning of its expected sixty-year lifetime. The cost of fueling the IFR the first time is a capital cost of the reactor—that is, part of the initial cost of the reactor. The IFR would not be refueled periodically like a thermal reactor; therefore, it has no refueling cost. The quantity of waste produced by the IFR is so small that the waste can be stored on-site until the plant has to be refurbished or removed.[16]

Chapter Two

What Can an Integral Fast Reactor Do?

The Integral Fast Reactor (IFR) and the Secure Transportable Autonomous Reactor-Liquid Metal (STAR-LM) can:

- generate electricity;

- provide direct heat or electricity needed to heat oil shale and power the equipment needed to extract the oil and natural gas or to desalinate sea water;

- create hydrogen gas by electrolysis.

The STAR reactor is also liquid-metal cooled, so the suffix "LM" is added—STAR-LM. The STAR reactor has all of the qualities of the IFR, as described here, and some worthwhile improvements to the IFR.[1] One of the improvements is a gas turbine recompression Brayton cycle that uses supercritical carbon dioxide as the working fluid instead water.[2] Supercritical carbon dioxide is carbon dioxide at its critical temperature of 31.1° Celsius and its pressure held at 73 atmospheres. Supercritical carbon dioxide expands to fill its container like a gas but retains a density like that of a liquid. To produce hydrogen and electricity, all that is necessary is to add a water supply and hydrogen storage facilities to the IFR or STAR-LM.[1a]

For our purposes, we can use the names IFR and STAR-LM interchangeably. To simplify future discussions, you may safely assume that the

STAR-LM is at least as capable as the IFR with regard to creating electricity and hydrogen.

Generate Electricity

The IFR can provide the heat energy to create the steam to drive steam turbine generators. The electricity produced by the generators can be used to run factories, light homes and streets, heat oil shale—all of the things for which we currently use electricity, and all of the things for which we will need electricity in the future.

Extract Oil from Oil Shale

Extracting millions of barrels of oil daily from oil shale is the essence of phase one of the Twelve-Years-to-Hydrogen Plan. A fact not well publicized in the United States is that there are, by one estimate, about 1.2 trillion barrels of oil bound up in oil shale in the West-Central United States. Another estimate puts this figure at 800 billion barrels[3, 6] By either estimate, this is an amount of oil greater than all of the oil ever found to date in Saudi Arabia. To help visualize this amount of oil, 800 billion barrels would fill 640,000 supertankers carrying 1.25 million barrels each. As previously mentioned, however, there is one big problem with oil shale: either America will have to wait 100 million years for nature to convert the oil in the shale into light sweet crude and natural gas, or we can heat the oil shale now to drive out the oil and natural gas. To date, two methods have been used to heat and extract oil and natural gas from oil shale.

The first method requires digging the oil shale out of the ground, heating it in a vacuum to drive out the oil and natural gas, then disposing of the waste. Some of the waste products using this approach are carcinogenic, so the disposal part of this process is a serious problem.

The second method is the in situ conversion process developed by a major oil company. The in situ method requires heating the oil shale in place, using heating elements placed 1,000 to 2,000 feet into the ground.

Using the in situ conversion process, the oil shale will be heated for a period of approximately three or four years to drive the oil and natural gas out of the oil shale. After the prescribed heating period, the oil and natural gas will be extracted from beneath the surface in the usual manner: by drilling wells. A 2006 experiment conducted by a major oil company, as part of Project Mahogany, produced over 1,500 barrels of light sweet crude using the in situ conversion process.[3]

The methods currently used to heat the oil shale are inefficient. It takes an amount of energy equal to the energy in one barrel of oil to heat the oil shale enough to drive out three to three and one-half barrels of oil—one barrel in, three-plus barrels out. Because the IFR can generate the needed heat directly or, if need be, generate electricity that can subsequently be used to create heat, the IFR could greatly reduce the cost of getting oil from oil shale. After all, we already have an abundance of low-cost nuclear waste fuel (plutonium) for the IFRs. As of December, 2005, the United States had 53,440 metric tons of spent nuclear fuel in 125 storage sites, and about 2,550 metric tons are being added annually.[4]

The completed environmental study of the Mahogany shale oil extraction project indicated that the only significant environmental impact of the in situ conversion process is on groundwater. To prevent groundwater contamination, a "freeze wall" would be created around the area of oil shale being heated. Freeze walls have been used for over a hundred years for this purpose and have been highly effective.[5, 7]

The in situ process for extracting oil from oil shale might sound like the answer to most of the United States' energy problems, but it is not. No matter how much oil or coal is extracted from the Earth, we are still stuck with the fossil fuel pollution that can eventually destroy the ecological balance of our planet. If we humans continue to maintain only fossil fuel—based economies, we will probably cease to exist. Ultimately, we will need to start using hydrogen as our primary fuel—and we will need to start very soon. Even with the use of hydrogen as our primary fuel, there still may be no guarantee of immortality for the atmosphere of planet Earth.

As an interim measure, IFRs could be used to enable the economical extraction of oil from oil shale. This interim measure alone would make the United States energy independent with respect to petroleum-based fuels for about one hundred years if no effort were made to transition to a hydrogen fuel infrastructure. Illustrations of in-situ heating methodology for extracting oil and natural gas by heating the oil shale in place rather than using strip-mining techniques can be found on the Shell Web site.[5] The freeze wall required to protect the groundwater would be created outside of the perimeter of the area being heated. Again, the IFRs would be instrumental in facilitating the in situ conversion process by providing electricity required for the refrigeration that would be used to create the freeze wall.[5] The sodium-cooled IFRs will probably be the only IFRs fully developed soon enough for implementation, but they would be well suited for use in the shale fields.

Successful, long-term extraction of a high volume of oil—10 to 15 million barrels per day—from oil shale would relieve public pressure on America's bureaucrats and forestall the United States' petroleum energy crisis for about a century[6]. However, without public pressure on the bureaucrats and legislated hydrogen production requirements and fuel taxes, the oil shale solution would be the only solution attained.

Natural gas is also produced in the process of getting oil from oil shale. This may seem to be a nice bonus, but in the long term, it is not. Burning any fossil fuel produces carbon dioxide as one of its many products of combustion.

The richest oil shale deposits in the United States are located in Wyoming, Utah, and Colorado.[6] Approximately 70 percent of all the oil shale deposits are on federal lands. As part of a larger agreement, the oil and gas resources on these federal lands should be made available without charge to the oil companies that agree to invest significantly in the oil shale fields.

Some of the oil companies could be producer companies that extract oil and gas from the shale. Other companies could be processor companies that process the oil into fuels and distribute the natural gas. There surely would be enough producing and processing requirements to keep the United States' oil oligopoly occupied.

New Oil Refineries

Oil companies are going to have to make some adjustments as the shale oil fields are brought into production. For example, most oil refineries in the United States are now located on the coasts, the exceptions being a few in the Midwest. Exploitation of the shale fields might necessitate locating some oil refineries near or in the states of Wyoming, Utah, and Colorado. Since oil refineries as we now know them are environmentally ugly at best, perhaps it would be possible to construct new refineries that employ more environmentally friendly technologies. Instead of reforming the natural gas recovered from the shale to provide the hydrogen needed to process the oil, perhaps one or two of the IFRs could be of the type that produces both electricity and hydrogen. The IFR-produced hydrogen would not result in greenhouse gas production.

The Fossil Fuel Business as Usual—Plus Tax

The Twelve-Years-to-Hydrogen Plan would mean a rebirth of the energy industries in the United States. Besides the construction of refineries, development of the oil shale deposits would probably require the construction of more pipelines for natural gas and pipelines to carry oil and hydrogen to the new refineries and to carry oil to the Gulf Coast oil export facilities. Essentially, oil from the oil shale fields would result in the same highly polluting fossil fuel business that we know today. Realistically, with the proper level of effort, development of the shale fields should be completed in six years. This would include a concurrent environmental impact study of the pipelines, refineries, shale heating, and groundwater pollution testing.

You might ask why oil would be needed if the United States has a hydrogen fuel infrastructure. The answer is that for many years we will still need many of the products that are derived from oil, like plastics, lubricants, and paving materials. Besides, the United States will need twenty to thirty years to wean itself from fossil fuels, so we would still need gasoline and fuel oil, but in continuously decreasing amounts.

Now for the Hydrogen

Phase two of the Twelve-Years-to-Hydrogen Plan entails creating the infrastructure to produce millions of gallons of hydrogen daily. Once the oil shale fields are beginning to heat, then production and installation of the remaining 480 to 490 modular IFRs should begin. Within the first six years of the plan, the oil shale fields and other domestic oil sources will be producing all of the oil needed domestically, so the focus of attention can then be the technologies needed to develop the hydrogen fuel infrastructure. Care must be taken that once the oil shale fields are developed, the nation does not then become complacent with respect to the environment—development of the hydrogen fuel infrastructure must proceed. To ensure that development of the hydrogen fuel economy does progress, a fossil fuel tax must be legislated at the beginning of the Plan that will result in pricing fossil-based vehicle fuels at an amount equal to 1.25 to 1.5 times the expected pump price of a hydrogen fuel energy equivalent. This should produce sufficient incentive for the nation to complete the hydrogen-based aspects of the Plan. The fossil fuel taxes would be in effect for the duration of the Plan and until hydrogen replaces gasoline as the dominant transportation fuel.

Generate Hydrogen Gas

Current light water reactors, because they typically operate at temperatures under 350°C, are not considered efficient hydrogen producers.[8] Integral fast reactors will reach temperatures high enough (700° to 900°C) to efficiently produce both hydrogen and electricity. The Integral fast reactor's electricity output can also be used for the electrolysis of water. More efficient hydrogen production may be attained by thermochemical splitting of water, also known as electrolysis of high-temperature steam. Another possibility is the use of nuclear energy as the source of heat for steam-methane reforming. However, this process also releases carbon dioxide into the atmosphere, whereas electrolysis of water and steam releases no carbon dioxide.

Efficient water-splitting and steam methane reforming processes all require temperatures above 700°C. Advanced reactors such as the advanced liquid-metal-cooled reactors can achieve the required high temperatures. The copious amounts of hydrogen produced will be pressurized by revolutionary electrochemical hydrogen compressors and liquefied by refrigeration units powered by IFR-generated electricity. Alternatively, the hydrogen might simply be compressed and then stored in high-pressure underground tanks. Some of the oxygen resulting from electrolysis should be liquefied and stored for industrial and medical purposes. However, most of the oxygen probably would be released back into the atmosphere. Each day, one 1-gigawatt (GWe) IFR could produce a quantity of hydrogen having an energy equivalent to about 1.025 million gallons of gasoline (or 4.1 million gallons of liquid hydrogen) and provide sufficient electrical power to compress and liquify the hydrogen. With 500 such IFRs in operation initially, the energy equivalent of approximately 512 million gallons of gasoline could be produced daily. This volume of hydrogen would require electrolyzing nearly five trillion gallons of water daily. IFRs have the potential to reliably create enough hydrogen to satisfy the needs of

America's hydrogen fuel infrastructure. If more hydrogen is needed, more reactors can be built.

Hydrogen Can Be Safe

We have discussed and described the energy-producing potential of the IFR. There is one more aspect of hydrogen that needs to be discussed—safety. Consider these facts about hydrogen:

- Hydrogen does not occur free in nature in useful quantities.

- Hydrogen burns without producing a visible flame.

- Hydrogen byproducts of combustion are mainly water vapor.

- Hydrogen liquefies when pressurized and cooled to about minus 425° Fahrenheit.

- Hydrogen is very costly to distribute in a pipeline like the type of pipeline used for gasoline.

- Hydrogen is difficult to contain in any type of container.

- Hydrogen has a relatively low energy density—the energy in four gallons of liquid hydrogen equals the energy in one gallon of gasoline. However, a gallon of gasoline is about ten times heaver than a gallon of liquid hydrogen.

We must realize that handling hydrogen instead of gasoline will require some training and accommodation on the part of motorists. Fortunately, most of the unique devices needed to implement a hydrogen fuel infrastructure have already been invented and developed. For example, a major oil company and others have already developed and placed commercial hydrogen fuel dispensers into daily use.[9] In most respects, hydrogen can be made to be about as safe as gasoline. The Department of Energy already has an online course available for first responders to a hydrogen mishap.[10]

Hydrogen Distribution

Given the fact that it is currently very costly to distribute hydrogen in a conventional pipeline, it would seem logical to assume that distribution of hydrogen to service stations would also be difficult. But NASA's space shuttle program has helped solve some of the problems associated with hydrogen fuel distribution. It may come as a surprise that liquid hydrogen fuel for the space shuttle is probably manufactured in Texas or Florida and transported to the launch site in special tanks. The space shuttle uses hydrogen as a fuel because it is light in weight. The liquid hydrogen is carried in an external fuel tank, which is jettisoned shortly after launch.

Initially, specially designed tank trucks could easily transport liquid hydrogen within a 150-to 200-mile radius of hydrogen-producing IFR locations. The problem of transporting hydrogen will be greatly simplified through local production of hydrogen at 500 or more widely dispersed IFR locations.[9] Where required, several IFR modules can be concentrated in one location to achieve hydrogen production economies of scale. Eventually, hydrogen pipelines could be constructed, where volume considerations warrant, at a cost of about one million dollars per mile. The high costs are due to the special treatment of the pipes that enables them to endure the severe thermal stress.[9]

Hydrogen Service Stations

There are about 176,000 service stations in the United States. Each service station would eventually have to be able to deliver the amount of hydrogen-derived energy needed to perform the same transportation functions as are currently being performed by the energy derived from gasoline. During a transitional period, however, the service stations will have to be able to meet the demand for both gasoline and hydrogen.[11]

Hundreds of Millions of Gallons of Hydrogen

Currently, the United States consumes about 409 million gallons of gasoline daily.[12] The 384.7 million gallons of gasoline shown in note 12 is a 2005 amount; 409 million gallons has been extrapolated from the 2005 amount. So, how much hydrogen would the United States need each day to replace the 409 million gallons of gasoline? If you assume that the hydrogen is going to be consumed by fuel cells, then the United States would eventually require about 613.5 million gallons of liquid hydrogen (347.9 million pounds, or 173,927 tons) daily, assuming the energy in four gallons of hydrogen is equal to the energy in one gallon of gasoline. The reason we would not need four times as many gallons of hydrogen to replace the 409 million gallons of gasoline is based on the assumption that all of the hydrogen would be consumed in fuel cells, and that the fuel cell is more efficient than the internal combustion engine. However, this assumption about "consumption in fuel cells only" may not be totally correct. A standard internal combustion engine (ICE) in which we now burn gasoline can be converted to burn gaseous hydrogen. The internal combustion engine is about 18.75 percent efficient when it burns hydrogen. The following rough calculations are based on the fact that the *gasoline-*powered internal combustion engine is about 15 percent efficient when all is said and done ("from well to wheel").[13] The fuel cell is about 40 percent efficient, or 2.67 times more efficient than the gasoline-powered internal combustion engine and 2.13 times more efficient than the hydrogen-powered internal combustion engine.[14] The figures used here are "in the ballpark," but they should by no means be considered precise. For detailed information on the principles of fuel cell operation, please see the Ballard Power Web site.[15]

Cleanliness Will Come in Phases

Because 176,000 service stations will need to be outfitted to dispense about 613.5 million gallons of liquid hydrogen (or its gaseous equivalent) each day, there is little hope of converting the entire nation to hydrogen, en masse, on a given day. The phasing in of hydrogen fuel would probably begin on the West coast and then progress to the Gulf Coast, then upward through the Midwest, and finally to the East coast. Initially, all vehicles will probably not be powered by fuel cells. There is a good possibility that many vehicles will be powered by hydrogen-fueled internal combustion engines. Even though the internal combustion engine is only about 18.75 percent efficient if it is burning hydrogen, the overall level of pollution it produces will be much less.

Subsidies and Overlapping Fuel Supplies

Once hydrogen fuel is available throughout the country, then the count-down for a de facto hydrogen fuel infrastructure will begin. Fossil-fueled vehicles will be around for fifteen to thirty years after the hydrogen fuel infrastructure is in place. Exactly how long the fossil fuel guzzlers will be around depends on how heavily the government is willing to subsidize the manufacturers and consumers of hydrogen-powered vehicles. The first fuel cell—powered vehicles will be expensive in relative terms. As the quantity of fuel cell–powered vehicles increases, then economies of scale will be realized, the cost of fuel cell–powered vehicles will decrease, and the amount of the government subsidy paid to manufacturers and consumers can be reduced. I can hardly wait to see the first hydrogen-powered lawn-mower.

Chapter Three

What Happens When the United States Stops Buying Oil?

What happens when the United States stops buying oil from other countries? This question may bring a smile to your face. In how many instances in recent times has the United States been held hostage by oil-rich countries?

There Will Be Significant Economic Impacts

There are important ecopolitical concerns in considering the advent of a net fossil fuel–producing United States. Oil revenues in some Middle Eastern countries will begin to decline. However, instead of shearing sheep, spinning the wool into yarn, and weaving and selling oriental rugs, or doing something else productive, the average Middle-Easterner will remain self-employed or otherwise gainfully employed in industry. Bear in mind though, that is not uncommon for unemployment to exceed 20 percent in some Middle Eastern countries. A surprising statistic: the combined gross domestic product of all Arab countries in 2003, including oil exports, was less than that of Mexico—about $540 billion,[1] compared to about $10.988 trillion in annual gross domestic product for the United States the same year.[2]

What would happen if suddenly there were a sizeable reduction in oil revenues in the Middle Eastern countries supplying oil to the United States? One might speculate that the leaders of these countries would be looking for some source of revenue to replace the lost oil revenue. Where might they go to solve their revenue generation problems? They would probably go to those who possess the greatest body of expertise on capital generation through entrepreneurial endeavors: the United States, Europe, and other modern countries. As the United States consumes less foreign oil, its oil-related trade dependencies would begin to reverse, as would the currency flow and trade deficits.

What Will Be The Cost of What We Need?

The cost of constructing and bringing on line a sufficient quantity of IFRs is a nebulous number at this juncture. The U.S. Department of Energy estimates that each reactor, IFR, or STAR-LM will cost about one billion dollars.[3] The cost of building and testing the first modular, transportable IFR will certainly be several times the cost of building the second and subsequent units. By identifying manufacturable designs for the IFR, then modular design, manufacturing, assembly, and the economies of scale can be brought to bear, and the cost of the IFRs will be minimized. One such optimal design may be derived from the Secure Transportable Autonomous Reactor, Liquid Metal (STAR-LM).

The name of the STAR-LM has some important implications; specifically, the words secure, transportable, and autonomous. "Secure" refers to the fact that a STAR reactor is proliferation resistant. "Transportable" implies that the STAR reactor is modular and can be transported on a train or a truck. "Autonomous" refers to the fact that the reactor can increase or decrease its output based on energy demand, without human intervention. These terms cannot currently be applied to today's LWRs. The STAR-LM reactor can use either water or carbon dioxide as a working fluid.

For the purposes of this discussion, let us say each modular IFR will cost $1.5 billion to construct and bring on line—this is probably a reasonable estimate. As an initial target, let us assume 500 IFRs will have to be constructed, at a total cost of $750 billion. The other significant figure that is not readily available is the cost of operating an IFR. The Department of Energy will probably be the entity to estimate IFR operating costs. There will be no refueling cost for the IFR, because it is fueled only once, and the initial fuel lasts for the life of the reactor. The cost of the IFRs amounts to a lot of money, but what will be the real cost of *not* building them? Besides, this is only about 50 percent more than what the United States will probably squander on the Iraq War.

The next question is where to locate the IFRs. The first nine to twelve IFRs will be sodium-cooled IFRs constructed in the oil shale fields to begin heating of the oil shale. The location of the commercial thermal reactors (LWRs) will dictate where many of the remaining 490 or so IFRs are located, because energy demand—related demographics have already dictated where thermal reactors are built. The IFRs' primary function will be to furnish hydrogen to towns and cities within a 150-mile radius of the IFR sites. Any electricity produced by the IFRs that is not used for hydrogen production can be applied to the same portion of the national power grid that the LWRs are supplying with electricity.

How to Estimate the Costs

One of the anomalies I encountered in assembling this tome is the tremendous variation in the estimated resources required to establish a hydrogen fuel infrastructure in the United States. To get a truly accurate estimate of the resource requirements, experts from all of the core technologies that will be required to make the hydrogen fuel infrastructure a reality must be consulted. The range of estimates that I encountered in attempting obtain reliable hydrogen fuel infrastructure cost estimates to include in this book was shocking—840 billion dollars to 22 trillion dollars.[6] These figures are the estimates of the cost to produce approximately 150 to 170 million tons of hydrogen daily using any one of the various technologies like natural gas, wind, solar energy, nuclear power, biomass, or clean coal. For example, one estimate of the cost of using *nuclear* energy to produce the 170 million tons of hydrogen included the following factors:

- 240,000 tons of unenriched uranium will be required—this is five times today's global production.

- Two thousand 600-megawatt, next-generation nuclear power plants will be required; only 103 nuclear power plants (LWRs) operate in the United States today.

Clearly, the originator of the *nuclear* energy cost estimates:

- Did not consider the possibility that the IFR would be used for hydrogen generation, nor did the originator appear to know the IFR fuel cycle. As previously explained, the IFR will use waste plutonium for fuel, so we can disregard the cost of mining and milling 240,000 tons of unenriched uranium.

- Appears to be assuming that the reactors used to produce the hydrogen will be LWRs. If so, the assumption is incorrect. Light water reactors do not get hot enough to generate hydrogen efficiently.

- Assumes that two thousand nuclear reactors will be needed. If this were the case, then the total cost of the nuclear option would be well over two trillion dollars if LWRs were the type of reactor used.[6]

It must be emphasized that the information presented in this book is from many sources, and regardless of the source, the information was extremely difficult, if not impossible, to verify. Cost estimates for a hydrogen fuel infrastructure made in the absence of a plan for constructing a hydrogen fuel infrastructure are prone to gross inaccuracies. What follows in this chapter is a combination of "best available guesses" mingled with the verifiable facts.

The IFR used at Argonne National Laboratories or the STAR-LM can serve as the prototype IFR design. The University of Chicago and General Electric Corporation will probably do much of the IFR design. By performing this design work from the perspective that the first IFR will be the prototype for up to 1,000 transportable nuclear reactor kits, then economies of scale can prevail during IFR construction and installation.

Some Thoughts on Management

Based on current planning at the Department of Energy (DOE), about 10 percent of the nation's hydrogen will be furnished by IFR technology by the year 2020.[7] Bush's Hydrogen Fuel Initiative reportedly calls for replacing fossil fuels used in passenger cars with hydrogen by 2040. Most climatologists would agree that the hydrogen initiative planning needs to be accelerated by ten to twenty years—about 10 percent of the nation's hydrogen could be furnished by IFR technology by the year 2016. The Bush & Co. budgets will not be sufficient to accelerate hydrogen infrastructure development to this extent.

However, if the administration that takes office in 2009 will immediately begin to implement the two-phase Twelve-Years-to-Hydrogen Plan and a $0.75-per-gallon fossil fuel tax to finance the Plan, then fossil fuel independence and a hydrogen fuel infrastructure could be in place by 2022. Approximately $109 billion per year would be realized from the vehicle fuel tax. The $109 billion in fuel taxes would pay for about 90 percent of the cost of implementing both phases of the Plan. Additional federal revenue would come from restoring higher income tax rates on the top 1 percent of income earners.

The year preceding the first year of the Twelve-Years-to-Hydrogen Plan would be a preparation year for completing the administrative organizations and IFR designs, initializing manufacture of the IFR designs, perhaps getting the oil companies on board, and collecting the first year's fuel taxes to get operating capital. At the end of the twelve plan years, the "new" hydrogen fuel taxes would replace the taxes of the Plan. The new fuel taxes would subsidize the manufacture of fuel cell—powered cars and some national social needs.

The Twelve-Years-to-Hydrogen Plan *must* be a consortium of the U.S. government and private energy industry companies. For twelve years, the U.S. government will own and set standards for the national energy consortium. The U.S. government should fund the construction and startup of all IFRs and all other aspects of the Plan using predominately the proceeds of the fuel tax. Each IFR would be constructed and tested by the government, then turned over to private nuclear power reactor management organizations when it is time to activate the reactor. The Nuclear Regulatory Commission should determine the function each IFR is supposed to serve—that is, electricity or hydrogen production, or desalination to remove salt from sea water, or some combination of these functions. The Nuclear Regulatory Commission should also certify the reactor management organizations and their personnel. The private nuclear power reactor management organizations could be public utility companies that currently operate nuclear power reactors, or they could be new subsidiaries of oil companies.

A portion of the hydrogen and electricity revenues created by the IFRs (about 40 percent to 50 percent) would be repaid to the government annually for a period of twenty years (or one-third of the reactor lifecycle) as a payback for the IFR reactors and fuel. After twenty years of successful operation the IFRs operated by the private companies would retain all of the revenue from the IFRs, with the proviso that they would always have government oversight. This would leave about forty years for all of the profits on each reactor to be retained by the management companies, assuming a sixty-year reactor life cycle. The management companies would be responsible for refurbishing the IFRs, when needed, under government supervision.

Government funding would also be used to install hydrogen-dispensing systems at the 176,000 service stations. To recoup these funds, the gasoline produced from shale oil would remain taxed at the rate of $0.75 per gallon; hydrogen would be taxed at the rate of $0.15 per gallon. Using IFRs, the estimated delivered cost of a gallon of liquid hydrogen, including tax, would be about $0.50. The tax on hydrogen would be collected for a period of ten years from the date the dispensing equipment was installed. The hydrogen fuel tax would augment the taxes on all fuels during and after the Twelve-Years-to-Hydrogen Plan. The cost of fossil-based vehicle fuel plus taxes should not exceed 1.25 times the cost of the same energy equivalent of hydrogen, which should not exceed $1.25 per gallon—including tax. Since the cost of IFR nuclear fuel used to produce hydrogen and electricity is only the cost of handling, reprocessing, and transportation for the nuclear fuel and waste disposal about every 60 years, the actual nuclear fuel costs are a relatively minor expense. Bear in mind that if the hydrogen is consumed in a fuel cell—powered vehicle, the consumer will receive the benefit of a 267 percent increase in efficiency.

Vehicle fuel taxes should remain sufficiently high (a few cents per gallon) to provide six billion dollars annually for a national scholarship fund and to subsidize Social Security funds. It should be possible to sell about 25 million barrels of shale oil annually for five years to help build these funds. Much good (as well as harm) could be done with the proceeds of fossil fuel sales. Here is the advantage to the American people of having the federal government *retain ownership* of the oil shale land. The oil shale land would be leased to the oil companies for one dollar for a period of fifty years, with the proviso that all reasonable efforts immediately be made to extract a specified quota of the oil and natural gas from the oil shale using current in situ extraction technologies. To charge the oil companies a high leasing fee for the oil shale land would essentially be a tax that would reduce their profits unnecessarily. Oil companies would typically

pass on the cost of leasing fees to the consumer in the form of higher product prices.

Regrettably, there will have to be a government bureau to oversee the
IFR construction projects and IFR operations. Additionally, another government bureau must oversee the hydrogen distribution and dispensing
systems for the first twenty-five years.

Hydrogen-Burning Internal Combustion Engines and Fuel Cells

Internal combustion engines can be modified to burn gaseous hydrogen.
Internal combustion engines will be cheaper than fuel cells for the first two
or three years after the hydrogen fuel infrastructure implementation is
begun. This is because there already is a tremendous amount of plant
capacity for building the internal combustion engines, but the plants
required to build millions of fuel cell power units remain to be built, and
the construction costs of the fuel cell plants must be profitably distributed
over the fuel cell units produced. For this reason, the fuel cell manufacturers' startup costs must initially be subsidized by the government.

The primary incentive to accelerate development of the fuel cell is the
significant difference in efficiency of internal combustion engine versus
fuel cells—the fuel cell is now about 2.67 times more efficient. An emerg-
ing technology known as Parallel Path Magnetic Technology has the
potential to double the current efficiency levels of the fuel cell–powered
vehicle by increasing the efficiency of the electric motors used to drive the
fuel cell–powered vehicles and the generators used to generate the electricity used to create the hydrogen.[8]

We Will Need the Catalysts

At this time, the preferred catalyst for proton exchange membrane or poly-
mer electrolyte membrane (PEM) fuel cells is platinum. In the future, a
better catalyst may come along, but until it does, platinum is it. The
migration of platinum from the automotive catalytic converter, where it is

now used to neutralize pollutants in the exhaust gases of vehicles powered by fossil fuels, to the catalyst in a PEM fuel cell is not intuitive. When platinum is used as a catalyst, the platinum is not destroyed; it simply facilitates chemical reactions that reduce the toxicity of exhaust gases, in the case of the catalytic converter. In the case of the PEM fuel cell, platinum is responsible for the efficient functioning of the fuel cell. When a PEM fuel cell that is designed to power a motor vehicle is properly supplied with hydrogen, the fuel cell produces water and electricity. The electricity powers the vehicle, and the water (vapor) is released into the atmosphere. There are many other types of fuel cells, but the PEM fuel cell currently holds the most promise for powering vehicles; however, this could change. The best advice that can be given to any entity that plans to manufacture large quantities of PEM fuel cells is: hoard platinum![9]

Late in the evening of July 29, 2003, National Research Council (of Canada) Institute for Fuel Cell Innovation (NRC-IFCI) researchers achieved a key milestone in hydrogen fuel research that promises major benefits for the hydrogen fuel industries. Using a prototype electrochemical hydrogen compressor, researchers compressed hydrogen to 5,000 pounds per square inch (psi). The prototype solid-state, multi-stage compressor incorporates a series of membrane-electrode assemblies (MEAs) similar to those used in proton exchange membrane fuel cells. It operates with the application of an electrical potential across the MEAs to affect the transfer of hydrogen with progressively increasing output pressure from stage to stage.

This non-mechanical compressor also addresses a number of shortfalls inherent to traditional hydrogen compressors. The electrochemical hydrogen compressor, for instance, has no moving parts and, therefore, addresses conventional hydrogen compressor issues such as wear and tear on parts, noise, and intense energy use. The electrochemical hydrogen compressor is also significantly more compact, making it adaptable to a range of situations. The Canadian researchers have now achieved their

2004 goal of a hydrogen compression pressure of 10,000 psi using the electrochemical hydrogen compressor.[10]

Innovations such as this beg the question: if there were a concerted, accelerated effort, how soon could we move these new technologies into the commercial arena? For the time being, the operative words will have to be "concerted effort." Somehow, the innovators in the hydrogen technology industries need to perform their efforts in concert in order to eliminate duplication of effort and wasting of resources. The Twelve-Years-to-Hydrogen Plan would provide the needed vision to help focus hydrogen fuel infrastructure resources and development efforts.

Improved Health for All Living Creatures

It is estimated that hundreds of thousands of people die each year from ailments caused by fossil fuel byproducts of combustion that escape into the atmosphere, and it is estimated that tens of billions of dollars are spent annually for the medical care of these people before they die. When the hydrogen fuel infrastructure matures, it is believed that most asthma suffers will be thankful, and that the incidence of asthma in our society will decrease significantly.[10a] It will also be interesting to track the incidence of lung cancer in highly polluted areas like the Los Angeles Basin, once the hydrogen fuel infrastructure is realized.[11]

Every year, oil spills on the high seas cause the deaths of millions of living creatures throughout the world. Companies that transport oil in ships obviously hold profits in much higher esteem than they hold the state of the world's ecology. Annually, billions of dollars are spent trying to clean up oil spills, but the results of these efforts are marginal, at best. Passing on to consumers the cost of trying to clean up the massive oil spills thereby decreases the overall efficiency of our society.

Chapter Four

The Twelve-Years-to-Hydrogen Plan

To arrive at this point, we have discussed the inventions, innovations, devices, technologies, processes, and policies that *can* lead the United States to oil independence and a hydrogen fuel infrastructure within twelve years of some future start date. What is truly lacking with respect to the Twelve-Years-to-Hydrogen Plan is a national leadership with the intestinal fortitude to legislate and fund the Twelve-Years-to-Hydrogen Plan.

If we were to start the Twelve-Years-to-Hydrogen Plan today, we would probably start by linking together the functions of the following ten Plan elements that will constitute the United States' energy future:

- Integral Fast Reactor (IFR) design

- extraction of oil from the oil shale fields

- electrolysis of water methodologies

- electrochemical hydrogen compressor technologies

- hydrogen storage and distribution systems

- hydrogen-powered automobile and truck technology

- hydrogen dispensing systems

- proton exchange membrane (PEM) fuel cell technology

- improvements in Parallel Path Magnetic Technology

- taxes and advertising to shape economic attitudes and behavior.

These are the essential elements of the Twelve-Years-to-Hydrogen Plan. The sequence with which each of these essential elements is added to the Plan is critical to the Plan's success.

In the Beginning, Taxes and Scholarships

There will be one preparation year preceding the start day for the Twelve-Years-to-Hydrogen Plan. During this preparation year, the critical path planning charts will be developed and initial projects started. While this may seem like insufficient time to prepare and to subsequently accomplish the Plan, it must be remembered that the preparation year and the Plan implementation will be labor intensive, and a quantity of employees must be hired that is commensurate with the level of effort. Based on the quantity and complexity of the tasks that must be accomplished to implement the Plan, I estimate that 50,000 to 100,000 people with appropriate skills will initially be required. The Plan technology-of-focus for these employees will shift as the Plan evolves. For example, initially, the IFR technology will be the technology-of-focus, then the in situ extraction process will be the technology-of-focus, etc. The $0.75-per-gallon fuel tax must go into effect on the first day of the preparation year. This tax will provide about $109 billion dollars of operating funds for the preparation year and the following twelve years of the Plan. Additional money, if needed, will be provided from the government's general fund as required, with an emergency priority.

Two benefits that will immediately be available to all eligible American citizens and funded by the $0.75-per-gallon fuel tax are:

• Three billion dollars will be made available annually for a national four-year scholarship fund. Students will be selected on the basis of high school grade-point average only; no entrance tests will be used.

• Three billion dollars will be contributed to the Social Security fund annually.

The reason for the scholarship fund is that institutions of higher learning will be given incentives to develop curricula for and begin granting degrees in IFR technology, fuel cell technology, cryogenics and liquid gas storage technologies, and electric motor and magnetics technologies. Development of the curricula will occur during the preparatory year before the Plan actually starts.

The Social Security fund will receive contributions as compensation to retirees for the years of "skimming" from this fund by our politicians. The six billion dollars required for these funds will come from the fuel taxes during the preparatory year and all twelve years of the Plan. After the pipeline to the Gulf Coast is completed and oil exports from the shale fields begin, then the dollars received each year for exported oil will supplement fuel taxes and be used for these funds through the last year of the Plan.

Much of the 53,000 tons of fuel for the IFRs is currently in "storage" in 125 spent fuel storage sites.[1] The costs for the IFRs will be the manufacturing and construction costs, the initial fuel processing and transportation costs associated with the first fueling of each IFR, and the ongoing operation and maintenance costs. Because the actual cost of IFR fuel would be very low, relative to the cost of fossil fuels, a modest electricity tax could be applied on a per-kilowatt-hour basis to help fund a national health care plan. A tax of one quarter of a cent per kilowatt hour would provide over $100 billion for a national health care plan. Even with this electricity tax, the cost of electricity to the consumer would still be only a fraction of what it is today.

The IFR plants must be developed first. The final plans for the IFR plant must include the designed-in ability to mass produce IFR plants using modular manufacturing and assembly techniques. The STAR-LM reactor plants could meet this requirement. There should be up to three IFR designs: one designed to produce only electricity, one designed to desalinate seawater, and one designed to produce hydrogen. Certain aspects of the IFR project were revived by Bush & Co., so it should take only a modest level of effort to make operational IFRs or STAR-LMs, and to begin the reactor site preparation. The initial nine to twelve reactors must be installed and operational by the end of the second project year. These first reactors will probably be the sodium-cooled IFRs, because they are the closest to being ready for deployment. The first IFR plant assembly and plant operation teams will need to be fully trained and ready to install the first nine to twelve reactors when they are delivered to the shale oil fields. All oil shale heating electrodes and electrical power preparations for the shale fields, refineries, and oil and natural gas pipelines must be under construction for each oil shale field undergoing development. Just building and bringing on line nine to twelve IFRs in two years is a tremendous task, but such a task does allude to the scope of Twelve-Years-to-Hydrogen Plan.

The first sodium-cooled IFRs will be erected at prescribed locations in all three of the main oil shale fields, as designated by the oil company with the most experience with in situ shale oil extraction. Two of the shale field IFRs should be configured to produce hydrogen as well as electricity. This hydrogen would be used where needed in the shale oil refining processes in lieu of the carbon dioxide—producing process of reforming the natural gas extracted from the shale fields.

The primary objective during the first two years of the Plan is to get the IFRs for the oil shale field heaters on line, begin heating the oil shale fields, begin chilling the freeze walls, and begin construction of the new refineries and associated pipelines. The overall objective at the end of the sixth year of the Plan is to make the shale oil fields productive (10 to 15 million barrels per day), to put the pipelines and refineries on line, and to commence producing distillate fuels—gasoline and fuel oil. Also by the end of the sixth year, construction and installation of as many of the remaining hydrogen-producing IFRs as possible should be underway. The detailed schedules for implementation of the Plan would be done using the Critical Path Method of project planning. The planning objective of the Twelve-Years-to-Hydrogen Plan is to focus the effort on each of the core technologies at a point in time such that the technology in question is ready for implementation when it is needed.

The management approach used should mimic the nuclear submarine projects. A U. S. Navy admiral (submarine qualified) should be the project manager for IFR production, installation, and operation; Navy admirals have exceptional knowledge of things nuclear and great familiarity with the Critical Path planning method. All facets of the Plan have to be "crashed," or accelerated, to the maximum extent allowed by safety constraints.

Military personnel will be required to augment the Plan workforce. All military persons working on the Plan must be American citizens and have a Secret clearance. Civilian personnel working on the IFRs must be American citizens and have a nuclear facility clearance or equivalent. After working in the Plan workforce for two years, active duty military personnel, both reserve and regular, will be eligible for the national four-year scholarship fund educational benefit. If they request to participate and receive approval, they will be automatically admitted to a university to ensure they have an opportunity to participate.

The Core Technologies

Within the set of ten essential Plan elements are the core technologies that must be fully developed by the time they are due to be implemented in the Plan. The core technologies include the following:

- Integral Fast Reactor (IFR)

- electrochemical hydrogen compressor

- hydrogen distribution system

- hydrogen dispensing system

- proton exchange membrane (PEM) fuel cell technology

- Parallel Path Magnetic Technologies

These core technologies are the ones that will be responsible for supporting the national transportation fleet.

How the Core Technologies Are Used

The role of the IFR in the Twelve-Years-to-Hydrogen Plan is the most critical of all the core technologies. The heat and electricity the IFRs produce for the oil shale fields and the shale field refineries are required to be in place and fully operational prior to proceeding to the second phase of the Plan—the hydrogen production phase. The oil shale fields must be in full production at the end of the sixth year of the Plan. Once the oil shale fields are in full production, then construction and activation of the remaining IFR plants designed for hydrogen production, desalination, and electricity production should be in the process of being manufactured, constructed, fueled, and tested.

The hydrogen-producing IFRs will require special facilities for compressing, liquifying, and storing liquid and gaseous hydrogen underground. The electrolysis of water will make gaseous hydrogen available to the IFR plant for compression and storage.[2] Alternatively, the more efficient process of electrolyzing steam can be used.[3] To liquify the gaseous hydrogen, the electrochemical hydrogen compressor will be needed to perform the desired level of compression of the gaseous hydrogen.[4] The electricity produced by the hydrogen production plant will also be used to power refrigeration units that will cool the compressed hydrogen down to a temperature where it becomes a liquid: -253°C, or about 20° Kelvin.[5] The electrochemical hydrogen compressor, because it has no moving parts, is significantly more efficient than a conventional gas compressor.

Once sufficient numbers of hydrogen-producing IFRs are present in the Western United States, then quantities of liquid hydrogen can begin to be distributed to hydrogen service stations.[6] Distribution of hydrogen will require hydrogen-powered tank trucks to be operational, and liquid hydrogen transfer equipment will have been perfected and put in place.[6] When it comes to the hydrogen-producing IFRs, it is necessary to understand that any IFR plant designed to produce electricity can be expanded to produce hydrogen as well, with the proviso that there is a steady supply of water to electrolyze and facilities to store the hydrogen produced. Any IFR plant designed to produce hydrogen can easily divert its electricity to the national power grid, if so designed.

When liquid hydrogen is available to a majority of consumers in the Western United States, then the vehicle manufacturers will be required to furnish hydrogen-powered vehicles to the Western United States. This, in turn, will require the fuel cell manufacturers to have furnished the American automobile manufacturers with proton exchange membrane (PEM) fuel cells. Converting the electricity from the fuel cell to a vehicular locomotive force will be done using an electric motor whose performance has been enhanced by parallel path magnetic technology.[7] This newly demonstrated magnetic technology reportedly increases the efficiency of electric motors by a significant amount. How much the parallel path magnetic technology can increase the efficiency of a fuel cell–powered automobile or truck remains to be seen.

Chapter Five

Current Status of the Core Technologies

Once again, the core technologies I currently consider essential to the United States' forthcoming hydrogen fuel infrastructure are the:

- Integral Fast Reactor (IFR);

- electrochemical hydrogen compressor;

- hydrogen distribution systems;

- hydrogen dispensing systems;

- proton exchange membrane (PEM) fuel cell technology;

- Parallel Path Magnetic Technologies.

These technologies can potentially form the foundation of the United States' hydrogen energy future, and the basis of Twelve-Years-to-Hydrogen Plan. The sequence in which each of these core technologies receives the focus of the development and implementation efforts is critical to the success of the Plan.

To see where we are going with our hydrogen-based future, it is essential that we know where we are with respect to the developmental state or status of each of the core technologies. The policies needed for the Plan should be accomplished in the legislation authorizing implementation of the Plan. The locations of the oil shale fields are immutable geographical facts. As the core technologies mature, they may advance and morph into completely new and different technologies that are not mentioned here, but will have, nonetheless, become technologically essential to the Twelve-Years-to-Hydrogen Plan.

State of the Integral Fast Reactor

As mentioned in Chapter 1, the Integral Fast Reactor (IFR) and the Secure Transportable Autonomous Reactor-Liquid Metal (STAR-LM) are revolutionary types of reactors. Before the IFR project was discontinued by the government, the IFR embodied a research and development effort to exploit natural, safety-related phenomena and other innovations into a nuclear reactor that would be a new and improved component of the next-generation nuclear power plants.

Recently, the Department of Energy (DOE) has been funded to move the United States toward a more practical way of using nuclear power to generate electricity. The current state of development of a project like the IFR or STAR-LM project is somewhat difficult to determine, because the DOE does not frequently update the documentation it makes public. However, DOE does estimate that the sodium-cooled IFR could be available for the production of hydrogen by the year 2015.[1] This is ten years before the lead-cooled, nuclear-powered source of hydrogen is expected to be commercially viable.[1] Nuclear-powered and renewable energy–powered hydrogen producing methods will eventually produce 100 percent of the United States' hydrogen requirement within the 2030 to 2040 timeframe.[2] It is during this timeframe that the United States' hydrogen economy is currently expected to be fully realized. The sodium-cooled IFR will be ready ten to fifteen years before any other nonpolluting, high-volume hydrogen producing method.[3] By virtue of this fact, some of the IFRs constructed will have to be sodium cooled.

Pros and Cons of the Very High Temperature Reactor

The Very High Temperature Reactor (VHTR) is receiving an undue amount of attention because of its unusually high operating temperature. A reason for all of the time and attention given to the VHTR is that water vapor (steam) can be electrolyzed more efficiently at high temperatures. The VHTR is, at most, 10 percent more efficient when performing electrolysis than the sodium-cooled IFR, which is up to 45 percent efficient.[4] This is the only real advantage a VHTR has over the IFR. The VHTR has the following disadvantages:

- It uses a "one-time-through" fuel cycle. As discussed in Chapter 1, a one-time-through nuclear fuel cycle is extremely wasteful of uranium.[4]

- It needs to be refueled with fresh uranium (U235) every five years or so, and it produces significant amounts of plutonium as waste.

- The VHTR is one of the hydrogen producing means that will not be ready until 2020.

I expect that the considerable VHTR development effort will probably prove to be a waste of time and resources, due to its inefficient use of fuel.[4]

Pros and Cons of the Sodium-Cooled IFR

In contrast to the VHTR, the sodium-cooled IFR has the following previously discussed advantages:

- The IFR is greater than 99 percent efficient.[5]

- It can be placed into service by 2015.[1]

- Using the sodium-cooled IFR or STAR-LM, all of the fuel we will require for our national electrical power and hydrogen needs for up to 1,000 years will be available in the form of waste plutonium from the spent fuel storage facilities and uranium that is already prepared for use.

- The sodium-cooled IFR and the STAR-LM are inherently safe, due to the design of their Advanced Liquid-Metal Reactors.[5]

The most significant disadvantage of the sodium-cooled IFR or STAR-LM is that the sodium-cooled IFR or STAR-LM does not produce hydrogen as efficiently as the VHTR because of the sodium-cooled reactor's lower operating temperature (500°C to 800°C).

Since about 20 percent of the IFRs constructed will primarily be used for electricity generation, the hydrogen production efficiency issue will not have a severe impact on the total number of sodium-cooled IFRs needed. While it is not wise to waste energy of any type, the lower efficiency with which the sodium-cooled IFR produces hydrogen will be more than offset by the fact that the sodium-cooled IFR's fuel efficiency is 99 percent overall, as opposed to less than 10 percent for the VHTR. In the final analysis, the sodium-cooled IFR—and, more optimally, the STAR-LM (lead cooled) because of its higher operating temperatures (700°C to 900°C)—are the best compromises for a national nuclear power source. Unfortunately, unless development of the STAR-LM reactor is accelerated, it will not be ready for commercial use until 2025.[1] Accelerating the development and deployment of the lead-cooled STAR-LM reactor will be one of the highest priority of the Plan.

More Than One Type of ALMR

Both the Secure Transportable Autonomous Reactor-Liquid Metal reactor and the Integral Fast Reactor use an ALMR. There are two or three types of ALMRs being considered for hydrogen production. The ALMR associated with the IFR at Argonne National Laboratories was a liquid sodium-cooled reactor. The sodium ALMR operates in the temperature range of 500°C to 800°C. An ALMR that uses liquid lead as its coolant is also a strong contender as a hydrogen producer, because the operating temperature of the liquid lead reactor is up to 200°C higher than the operating temperature of the liquid sodium reactor. The higher operating temperature can improve the efficiency with which the liquid lead ALMR performs electrolysis. Because liquid lead does not react with air, its reaction with water is much less violent, and its shielding qualities are unsurpassed by other coolant metals make the lead-cooled reactor an excellent choice.

Characteristics of six of the most promising nuclear reactor systems are shown in the following table.[1]

Table of Nuclear Reactor Systems, Expected Deployment Dates, and Operating Temperatures[1]

Nuclear Reactor System	Expected Deployment Date	Operating Temperature °C
Sodium-Cooled Fast Reactor	2015	500 to 800
Very High Temperature Reactor	2020	Greater than 1000
Gas-Cooled Fast Reactor	2025	750 to 950
Molten Salt Reactor	2025	Greater than 700
Supercritical Water-Cooled Reactor	2025	550
Lead-Cooled Fast Reactor	2025	550 to 900

An important distraction on the road to a hydrogen fuel infrastructure in the United States is collaborative efforts with other nations.[1] While the United Kingdom, France, Russia, Canada, Japan, and China have been doing research with Liquid-Metal reactors, it is unknown how much mutually beneficial information will be exchanged between the countries. The tendency during such collaborative efforts is to keep significant research findings close to the chest for the homeland. Another negative aspect of collaborative efforts in research is the "not invented here" syndrome—that is, "we" did not invent it, therefore it must be inferior. This situation usually stems from the desire to make the final product "all French," "all Russian," or "all American," etc.

The Oil Shale Fields

As mentioned in Chapter 2, a major oil company is experimenting with an in situ method of extracting oil and natural gas from oil shale. To nationalize the in situ conversion process will probably necessitate paying a royalty to this major oil company for exclusive rights to its patents related to the process. The in situ process involves embedding heating elements 1,000 to 2,000 feet deep down into the shale, and then heating the shale to a temperature of 700° Fahrenheit for about three or four years. A fundamental problem with the process is that the oil may contaminate the groundwater further down into the ground after its release from the shale. To solve this problem, the company has buried refrigeration pipes in a ring around the oil extraction site to form a "freeze wall" around the edges of the extraction site so the edges will remain frozen solid and keep the liquid oil from contaminating the groundwater. Freeze walls have been used in this manner for over one hundred years.[6]

The company's experimental process requires a great deal of energy, but in the final analysis, it produces about three and one-half times as much energy as it takes to extract the oil and gas. In other words, the Energy Return on Energy Invested (EROEI) is low compared to conventional crude oil extraction methods. The heating process also produces natural gas, which, according to current thinking, can subsequently be used to produce heat to be used in the oil extraction process.

The length of time required to sufficiently heat the shale and the need for a freeze wall are both drawbacks to the in situ approach for extracting shale oil.

In the future, as stated in the Plan, electrical current to heat the oil shale and to power refrigeration systems for the freeze wall could be provided free of charge by the U.S. government's IFRs that are to be constructed in each of the three primary oil shale fields. The three primary fields can also be subdivided into several more fields that can be developed incrementally.[7]

Electrolysis of Water

The United States already uses over 9 million tons of hydrogen each year for industrial purposes, such as making fertilizer, refining petroleum, and launching the space shuttle. This is enough hydrogen to fuel about 34 million cars.[8] If hydrogen-powered vehicles are to become the norm, we will annually need at least 17 times more hydrogen. The challenge will be to produce hydrogen in an efficient, cost effective, and environmentally friendly way.

At present, 95 percent of America's hydrogen is produced from natural gas.[9] Through a process called steam methane reformation, high temperature and pressure break the methane (a hydrocarbon) into hydrogen and other gases, including carbon dioxide.

The Department of Energy projects that the nation's consumption of fossil fuels will continue to rise, increasing 32 percent by 2020. When burned, these carbon-based fuels will release millions of tons of carbon dioxide gas into the atmosphere; carbon dioxide gas will help trap the heat in the atmosphere and contribute to global warming.[10] This fact should help justify accelerating the development of the hydrogen fuel infrastructure.

Over the next ten or twenty years, it is estimated that fossil fuels most likely will continue to be the main feedstock for the hydrogen economy. The problem is, using natural gas to make hydrogen does not solve the pollution problem—it just adds to the problem by releasing more carbon dioxide into the atmosphere. Capturing the carbon dioxide and trapping it underground (sequestration) would make the process more environmentally friendly in the short term.

Most of the remainder of today's hydrogen is made by electrically splitting water into its constituent elements, hydrogen and oxygen, through electrolysis. Fossil fuels currently generate more than 70 percent of the electrical power delivered to the national power grid. Producing hydrogen using electricity from the national power grid would, therefore, contribute a significant amount of greenhouse gases to the atmosphere. If solar, wind, other renewable resources, or an IFR generate the electricity, hydrogen could be produced without any carbon emissions.

Water is generally available throughout the United States. When 2.3 gallons (9 liters) of water is turned to hydrogen and oxygen using electrolysis, the hydrogen energy equivalent of one gallon (3.785 liters) of gasoline is liberated. If the hydrogen from 2.3 gallons of water were liquefied, there would be about four gallons of liquid hydrogen. When burned or consumed by a fuel cell, hydrogen combines with oxygen in the atmosphere and produces water as the primary exhaust product. With combined cycle efficiencies (electrolysis + fuel cell) of over 60 percent, plus the 30 percent energy savings from the electrochemical hydrogen compressor and the potential efficiency gain through parallel path magnetic technology, I estimate we will likely exceed the overall efficiency of all fossil fuel extraction methods and any form of national electrical grid power except fourth-generation nuclear reactors. All of this should give us good reasons to pursue and further promote the nuclear-powered electrolysis process as the predominant source of hydrogen.

The Electrochemical Hydrogen Compressor

At room temperature and pressure, hydrogen's density is so low that it contains less than 1/300 the energy in an equivalent volume of gasoline. In order to fit into a reasonably sized storage tank, hydrogen has to be somehow squeezed into a denser form. Mechanical compressors that are used to compress hydrogen are very unreliable and consume large amounts of energy. In fact, a mechanical hydrogen compressor consumes an amount of energy roughly equal to one-third of the amount of energy held in the hydrogen it compresses.[12]

Enter the electrochemical hydrogen compressor—a truly amazing device. *Without any moving parts*, it can compress hydrogen into a container to a pressure of 10,000 pounds per square inch (psi), or more. This device is so critical to the hydrogen fuel infrastructure because the electro-chemical hydrogen compressor will save a tremendous amount of energy when liquefying large quantities of hydrogen. Before liquefying hydrogen, it usually needs to be compressed.[12] The most significant level of develop-ment for the electrochemical hydrogen compressor was the 10,000-psi level. 10,000 psi is the amount of pressure industry experts believe will be required to give a fuel cell–powered car a "run time" of 300 miles between refills when using gaseous hydrogen as the fuel for a PEM fuel cell. This pressure would make the fuel cell–powered car's range competitive with that of internal combustion–engined cars.

Some prototype hydrogen-powered vehicles really do use tanks of room-temperature hydrogen compressed to 10,000 psi. The Sequel, which GM unveiled in January 2005, carries eight kilograms of hydrogen com-pressed to 10,000 psi—reportedly enough to power the vehicle for 300 miles. Refueling with compressed hydrogen is relatively fast and simple. But even when compressed, hydrogen requires high-volume tanks. They take up four to five times as much space as a gasoline tank with an equiva-lent mileage range. However, because cars powered by fuel cells contain fewer mechanical parts, they have the available space needed to accommo-date bigger fuel tanks.[13]

The prototype solid-state, multi-stage electrochemical hydrogen com-pressor currently being developed by Canadian researchers at the National Research Council (NRC) of Canada incorporates a series of membrane-electrode assemblies (MEAs), similar to those used in proton exchange membrane fuel cells. The electrochemical hydrogen compressor operates by applying an electrical potential across the MEAs cause the transfer of hydrogen with increasing output pressure from stage to stage. The non-mechanical electrochemical hydrogen compressor addresses a number of reliability shortfalls inherent in conventional compressors because there are no moving parts, eliminating such issues as wear and tear, noise, and energy-use intensity. The electrochemical hydrogen compressor is also significantly more compact than the mechanical hydrogen compressor, making it adaptable to a wider range of situations.

The advantages of the electrochemical hydrogen compressor become very clear when you consider the compressor requirements of an IFR that produces hydrogen twenty-four hours per day, seven days per week. In such demanding circumstances, a mechanical compressor would pose significant maintenance and safety challenges. Given its design and the materials used, the Canadians' electrochemical hydrogen compressor presents a viable alternative. Whether electrolyzed hydrogen or a reformed traditional fuel like natural gas is used, the electrochemical hydrogen compressor will play a pivotal role in establishing a hydrogen fuel infrastructure.[14]

Current Hydrogen Distribution Technology

To be distributed, hydrogen must be temporarily stored and simultaneously transported to a location where it will be used to fuel a vehicle.[15] Hydrogen transportation and storage technology is well proven: for decades, NASA has used liquid hydrogen to power vehicles such as the space shuttle. Containers for gaseous or liquid hydrogen have many special requirements. Even in specially designed tanks, hydrogen is a difficult substance to move from place to place. Because of the temperature extremes, liquid hydrogen can embrittle steel and other metals, weakening them to the point of fracture. When chilled to a temperature near absolute zero, hydrogen gas turns into a liquid that contains one-quarter of the energy in an equivalent volume of gasoline. High-pressure and liquid hydrogen storage tanks are bulky, heavy, and expensive. However, high-pressure tanks should be considered a stopgap measure until materials can be developed that will allow us to store liquid hydrogen efficiently.[16]

Generally, hydrogen pipelines are located near the facilities that use large quantities of hydrogen—such as oil refineries. Currently, the longest hydrogen pipeline network in the world is a 1050-mile line in Texas.[17] Treating the metal pipes used in hydrogen pipelines to protect them from embrittlement due to temperature extremes and high pressure makes them expensive—about $1 million per mile. But once built, they are the cheapest way to deliver high volumes of hydrogen.

Alternatively, given the difficulty of transporting hydrogen, much of the hydrogen used today is created near the places where it is needed. This is what is done for roughly half of the thirty-six hydrogen fueling stations currently operating in the United States.[16] Four of the fueling stations rely on natural gas; the rest use electrolysis. In 2003, Honda introduced a Home Energy Station that performs steam reformation right in the owner's garage, but because natural gas is the feedstock, it still releases carbon dioxide into the atmosphere.[18] The IFR will be able to make the hydrogen close to the locations where it will be used, thereby reducing the Plan's initial hydrogen distribution infrastructure cost. Electrolysis produces the cleanest, purest hydrogen. This "clean" characteristic of hydrogen produced by electrolysis is extremely important in extending the life of PEM fuel cells. The more we find out about the desirable characteristics of the core technologies, the more obvious it becomes as to why each of the core technologies was selected.

Today's Hydrogen-Powered Vehicle Technology

While most of the applications for fuel cells are automotive, the Georgia Institute of Technology has done some interesting research with unmanned aerial vehicles. "Georgia Institute of Technology researchers have reportedly conducted successful test flights of a hydrogen-powered unmanned aircraft believed to be the largest to fly on a proton exchange membrane fuel cell using compressed hydrogen. The fuel-cell system that powers the 22-foot wingspan aircraft generates only 500 watts."[19]

The most straightforward approach to using hydrogen as a vehicle fuel is to burn hydrogen in a specially adapted internal combustion engine. Since little modification is required, these engines are relatively cheap, and they are about 25 percent more efficient when using hydrogen than the same internal combustion engine powered by gasoline. BMW built its first hydrogen internal-combustion engine back in the 1970s. BMW recently tested a hydrogen-fueled internal combustion engine at a speed of about 187 miles per hour.[20] Ford began production of a hydrogen internal combustion–engined shuttle bus in July 2006.[21]

At the time of this writing, "heavy-duty fuel cell power plants from Ballard Power Systems are powering the largest fleet of fuel cell buses in the world.[22] Over a two-year period, thirty Mercedes-Benz Citaro buses equipped with 205 kW Ballard fuel cell power plants will be in revenue service in nine cities in Europe as part of the European Fuel Cell Bus Project, and in Reykjavik, Iceland, as part of the Ecological City Transport System. Cities participating in the EU-funded project are: Amsterdam, Barcelona, Hamburg, London, Luxembourg, Madrid, Porto, Stockholm, and Stuttgart. In 2004, an additional six fuel cell buses entered service: three with the Sustainable Transport Energy Project in Perth, Western Australia, and three with the Santa Clara Valley Transportation Authority in California, USA. In 2005, another three buses began operation in Beijing, China. These demonstration buses are operating on regular transit routes in each city and will provide valuable information about the operation of fuel cell vehicles under real-world conditions. The bus fleets will also increase public awareness and aid in the further development of a hydrogen fuel infrastructure. As well, they will seed the market and provide thousands of people across Europe, California, Western Australia, and China with the opportunity to experience first-hand the clean, quiet, and comfortable ride of these zero-emission buses. Ballard Power Systems is the leading developer and manufacturer of proton exchange membrane fuel cells for transportation and power generation applications with over twenty years of experience."[25]

It is apparent that transportation companies throughout the world are being proactive adopters of hydrogen fueled vehicle technology.

Hydrogen Dispensing Systems

"Located within a Shell service station in Washington DC, a corporate visitor center interprets the importance of hydrogen fuel as a potential new fuel of the future. This is the first retail station in the United States with a hydrogen fuel dispenser, and Shell included a visitor center for educating the general public as well as government policymakers and think tank organizations within the District. The story is broken into three categories: a basic introduction of the project in Washington DC; "Why Hydrogen Fuel?" addresses concerns about the environment, the economy, and the safety of the fuel; and "Hydrogen Fuel, Here Today" discusses hydrogen fuel projects around the world and the development of hydrogen fuel vehicles. As Shell opens more hydrogen fuel dispensers around the world, the content developed for Washington DC will be localized and repeated at the respective sites."[23, 32]

"The drive to usher in a new hydrogen fuel infrastructure has taken a step closer to reality with the installation of the nation's first hydrogen pumps at a corner gas station. Hoping to show off the promise of a still-evolving technology, Shell Hydrogen and General Motors Corporation have unveiled two hydrogen dispensers at a busy Shell station just five miles from the Capitol, in Washington, DC. The project is not really a commercial venture. The dispensers will be used to fuel only six fuel cell–powered Opel Zafira minivans owned by GM. But the partners are hoping the availability of liquid and compressed hydrogen fuel at an otherwise typical gas station will intrigue energy policymakers and the driving public alike."[22]

Shell Hydrogen apparently believes that the combined hydrogen-gasoline retail outlet will give Shell customers an early introduction to an attractive motoring future. Government leaders have seized upon hydrogen as a potentially revolutionary fuel source that could help reduce urban smog, curb greenhouse gas emissions linked to climate change, and lessen the nation's dependence on foreign oil.

President Bush unveiled a $1.7 billion proposal in 2003, aimed at prodding hydrogen technology development.[24] In 2006, California Governor Arnold Schwarzenegger announced a plan aimed at installing 150 hydrogen filling stations in the state by 2010.[26]

"Estimates of just how much it would cost to build the necessary hydrogen distribution network vary widely. To get the effort rolling, the officials from Shell oil company have estimated the industry would have to spend perhaps $20 billion to install hydrogen facilities at 15 percent to 20 percent of U.S. gas stations and service 3 percent to 5 percent of the vehicles on the road. The technology, however, has its skeptics. Some Democrats, for example, have accused the Bush administration of focusing on long-range hydrogen technologies so as to divert attention from more pressing needs to diversify supplies and reduce fossil fuel consumption."[27]

Certain energy industry rivals allegedly believe that the cost of producing and distributing hydrogen would push up fuel costs to twice that of gasoline on a cents-per-mile-driven basis. What is missing is an explanation of how they expect the hydrogen will be produced. This is a classic example of referenceless, skeptical rhetoric that creates myths about a hydrogen fuel infrastructure. One must know what basis is used to estimate the alleged costs. When such skeptical, disparaging rhetoric is asserted to be true, it should be ignored until the purveyor of the rhetoric proves it to be correct.

Once again, it is the goal of this work to illuminate a path to fossil-based energy independence and a hydrogen fuel infrastructure in the United States by identifying the most promising technologies, and the sequence in which to employ them. This goal is summarized in the Twelve-Years-to-Hydrogen Plan. If this plan is properly executed, it will be 90 percent self funding and result in oil independence for the United States in six years, and a complete hydrogen fuel infrastructure in another six years: twelve years to hydrogen.

The small island country of Iceland wants to be the first hydrogen society in the world, a place where non-polluting renewable energy fuels everything from cars to homes and industry. An ambitious goal perhaps—but earlier, in 2004, the country's first public hydrogen filling station opened outside the capital city of Reykjavik. The city is one of ten in Europe that has pledged to build hydrogen filling stations to refuel thirty hydrogen-powered buses.[28]

Proton Exchange Membrane (PEM) Fuel Cell Technology

First invented in 1839, a fuel cell combines hydrogen and oxygen to generate electricity without any moving parts. Several different varieties of fuel cells exist, but only the proton exchange membrane (PEM) fuel cell is lightweight and responsive enough to be practical for vehicle use. Five or six years ago, the use of PEM fuel cells was considered cost prohibitive because, even in mass production, they were projected to cost about $36,000 each—primarily due to their platinum content. "Now, the current estimated cost of a PEM fuel cell is $67.00 per kilowatt of output. Eighty kilowatts is enough power to propel an automobile in a normal manner. Today, the cost of an 80-kilowatt PEM fuel cell is estimated to be $5,360."[31]

In another recent article, a company states they have developed a nickel-based compound, through the use of nanotechnology, that will reduce the cost of fuel cell catalysts by 75 percent.[31a]

Hydrogen is the universe's simplest atom: a single electron orbiting a single proton. In a proton exchange membrane fuel cell, incoming hydrogen gas is separated by a catalyst at the anode (ideally platinum) into protons and electrons. The protons pass directly through a proton exchange membrane (PEM), while electrons are forced through an external circuit, causing electric current to flow. When the protons and electrons meet at the cathode, they join with oxygen to form water vapor and produce heat, both of which are released into the atmosphere as exhaust.

A single PEM fuel cell produces just over 1 volt, so hundreds of cells are stacked together to supply enough voltage for typical applications. The PEM fuel cells used in NASA's Gemini flights in the 1960s are said to be the design of choice for fuel cell—powered cars, and other configurations are suited for applications ranging from laptop computers to electric power plants.

The catalyst, platinum, is a critical ingredient in the PEM fuel cell. Because of the cost of platinum, scientists are continually searching for alternative catalysts. Today, gasoline-powered cars use platinum as the catalyst in their catalytic converters. In view of the fact that platinum is a critical component in cars and light trucks, it would be prudent to accumulate a reserve of platinum. However, researchers also have recently created a platinum and nickel alloy that is reportedly superior to platinum alone as a catalyst—discoveries like this one can significantly reduce the cost of fuel cells and improve their efficiency.[29]

Parallel Path Magnetic Technology

Parallel Path Magnetic Technology is an advanced magnetic force control technology that is applicable to motors, rotary actuators, linear actuators, and generators. This technology uses permanent magnets controlled with a field coil in parallel magnetic circuits inside devices like motors and generators. Parallel Path Magnetic Technology is a simple but revolutionary concept that has been demonstrated in a wide variety of prototype devices. Devices that employ this technology are smaller, lighter, run cooler, and are more energy efficient than their conventional counterparts. As of this writing, there is no known, well-publicized commercial product that incorporates this technology.

Of the core technologies, Parallel Path Magnetic Technology will require the most research and development. The experimental data suggest that a substantial increase in the efficiency of the electric motor and the electric power generator is a possibility with Parallel Path Magnetic Tech-

nology. The ramifications of such an increase in electric motor and generator efficiency could reduce the nation's liquid hydrogen requirements by
at least 30 percent, and improve the efficiency of electrical generators. The
efficiency of the fuel cell—powered vehicle could potentially increase to
about four or five times the efficiency of the internal combustion engine.
This is the magnitude of efficiency improvement needed to make a hydrogen fuel infrastructure the most efficient of all possible options.[30]

Chapter Six

Where is the Planning for a Hydrogen Future?

After viewing some of the documentation on the Department of Energy's Web site, it would not be unfair to describe its current state of planning for oil independence and a hydrogen fuel infrastructure as being vague. The Department of Energy appears to be attempting to partner with many countries that might be able to contribute to the United States' endeavors, and vice versa. These organizations seem to provide almost every conceivable reason and almost every possible excuse to *not* develop a focused plan to achieve oil independence or a hydrogen fuel infrastructure with all dispatch.

Under the guise of attempting to find the most perfect solution to America's energy problems, the apparent tack taken by the Department of Energy to avoid perturbing American oil companies is to study every aspect of all possible approaches to solving the United States energy problems. This approach to studying the hydrogen fuel infrastructure options could last in perpetuity.

The Department of Energy is currently placing special emphasis on a type of reactor, the "Very High Temperature Reactor," that will waste a significant amount of nuclear fuel to achieve a few percentage points increase in the efficiency of electrolyzing water. Fuel economics dictate that the very high temperature reactor research just isn't worth the time and effort. This type of research is not logical when the IFR has the preferred "closed" reactor fuel cycle and fuel economics. One has to wonder about the motives for promoting the very high temperature reactor. The liquid-lead-cooled fast reactor, due to its capability to operate at higher temperatures, can achieve electrolysis efficiency approaching that of the very high temperature reactor at a much lower cost. A lead-cooled reactor, either an IFR or a STAR-LM reactor, will probably prove to be the most suitable reactor for the United States' near-future energy needs because of its numerous desirable qualities as a reactor coolant. In the final analysis, a few of the IFRs used in the Plan will probably have to be sodium-cooled reactors, due to the fact that they will be ready for commercial use by 2015, or possibly sooner.

It appears that the consensus fuel of the future is hydrogen. There has been a tremendous amount of research and development associated with hydrogen technology in the last few years. The core technologies can now be identified—these are the technologies that will receive the focus of the Twelve-Years-to-Hydrogen Plan. With all of the positive opinions and facts associated with hydrogen, what is the reason for all of the foot dragging? Why isn't the hydrogen fuel infrastructure being built? The answer is: a lack of both leadership and a basic plan that would focus research and development efforts on a desired outcome.

The gaps, if any, between the present-day states of the core technologies and their commercial viability will shrink much more rapidly if Congress funds and directs the Department of Energy to execute the Plan described here—or a similarly focused plan that will achieve the same goals in approximately the same amount of time. Initially, it will not be necessary for the hydrogen economy to be economically competitive with the present fossil fuel economy because there will be a fuel tax structure and government subsidies to defray the cost differences to the consumer. The oil shale fields will be brought into production as inexpensively as possible, but in a manner that is safe and environmentally friendly. The hydrogen economy will be created in a manner that will be less costly to consumers in the long term than the current fossil fuel economy.

The oil companies that want to share in the oil shale field developments should be informed that they would need to construct the pipelines and "clean" refineries as necessary to process the expected amount of oil. The oil shale land should be leased to the oil companies at no cost with the proviso that they will agree to government oversight for the life of the shale oil fields project, and that they will produce oil from the fields for as long as the fields are capable of producing oil.

By having a clearly defined plan of how the United States is going to become oil independent and establish a hydrogen fuel infrastructure, the Department of Energy, if so directed by Congress, will know to focus only on the core technologies rather than on every technology that could possibly ever have some practical value in a hydrogen fuel infrastructure. Unfocused, directionless research only results in collecting facts that have no immediately known purpose. I believe that we do not have the time to search for years to identify the ultimate optimum solution, of which there may be many of equal efficacy. Let us select the best combination of known technologies and begin to cure the world's oil addiction.

Summary

A set of core technologies has been identified here, and the developmental states of these technologies have been discussed and revealed as best possible from the sometimes conflicting data on the Internet. The identified core technologies for achieving oil independence and a hydrogen fuel infrastructure are:

- sodium-cooled Integral Fast Reactor (IFR) and the lead-cooled STAR-LM;

- in situ conversion process (oil shale fields);

- electrolysis of water;

- electrochemical hydrogen compressor;

- hydrogen distribution systems;

- hydrogen-powered automobile and truck technology;

- hydrogen dispensing systems;

- proton exchange membrane (PEM) fuel cell technology;

- Parallel Path Magnetic Technology.

Examples of the core technologies can be found in the public domain; for example, on the Internet. An in-depth review of the Web pages noted here will confirm that the science cited is well founded in empirical fact. While not a core technology, fuel tax incentives and advertising will be required to shape consumers' economic behavior throughout the course of the Plan.

So, where do we go next? How do we get from being aware that these technologies exist and being aware of what they can do for us, to implementing these technologies to the extent they can make the United States energy independent? What will it take to convince the dedicated energy capitalists that in some instances capitalism must be restrained to prevent some business sectors within an economy from being overwhelmed by the forces of avarice?

One logical answer is to have the Federal government create and fund a bill that fully supports all aspects of the Twelve-Years-to-Hydrogen Plan, and immediately begin implementing the Plan as it is described here. Implementing the Plan would have a synergistic effect by forcing focused development of the core technologies.

Another answer might be to have the Twelve-Years-to-Hydrogen Plan challenged by academia to flesh out the myriad details involved in a project that will totally restructure and temporarily nationalize the energy industry in the United States. Would such a challenge by academia obliterate any hope of implementing a plan, or would it excessively delay implementation of a plan? Can we afford to wait for academia at all?

It would seem that any action that delays implementation of the Plan should be immediately evaluated for its benefit to the Plan. Evidence appears to indicate that the global climate change we are currently witnessing is the greatest threat humankind has ever known; time is of the essence.

There appears to be a national consensus that oil independence for the United States is essential to its physical security. Oil independence is within the realm of possibility for the United States today; do we exploit this fact? Pollution-free energy is essential to preserving the atmosphere of our planet, and it too is within the realm of possibility today. Do we proceed to sequentially accomplish both phases of the Twelve-Years-to-

Hydrogen Plan? Or, do we contemplate the alternatives ad infinitum, as has been the apparent course of Bush & Co. and their predecessors, and imperil our planet and our country until there is no time left for alternatives?

The answer to any one of the forgoing questions is the same for all of the forgoing questions. The answer is: the citizens of the United States must create a political movement powerful enough to force our elected representatives to pass and fund a bill that mandates a plan equal to the Twelve-Years-to-Hydrogen Plan, and we must elect a president with the courage to sign the bill into law. Therefore, the only delay in implementing the Plan would be the time it takes for the electorate to create the movement, and the time for the movement to have the desired effect. To create such a movement will require educating the public on the feasibility and necessity of the Plan.

The government of the United States will not initiate a plan like the Twelve-Years-to-Hydrogen Plan of its own volition, because of the influence the energy corporations' lobbyists have on Congress. The United States' electorate *will have to cause* Congress to create and fund a bill for the Twelve-Years-to-Hydrogen Plan, if American energy independence is to become a reality.

Should Congress approve the Twelve-Years-to-Hydrogen Plan or its equivalent, then the United States will have more than enough carbon-free fuel to supply the nation with electrical power and hydrogen until the ultimate reactor is fully operational—the fusion reactor. Transitioning from the IFRs and STAR-LMs to fusion reactors should be much less traumatic than initially establishing a hydrogen fuel infrastructure.

Appendix

Conversion of units of weight and measure

To convert one unit of weight or measure to another, try this Web site:

http://convertworld.com

Useful Information about Hydrogen

- When 2.3 gallons (9 liters) of water is turned to hydrogen (through electrolysis), the hydrogen energy equivalent of one gallon (3.785 liters) of gasoline is liberated.

- Hydrogen has more energy per unit mass than other fuels (61,100 Btus per pound versus 20,900 BTUs per pound of gasoline). The problem with hydrogen is that it is much less dense (pounds per gallon) than other fuels. A gallon of gasoline has a mass of 6.0 pounds. The same gallon of liquid hydrogen only has a mass of 0.567 pounds, or roughly 9.45 percent of the mass of gasoline. Therefore, one gallon of gasoline yields 125,400 BTUs of energy, while a gallon of liquid hydrogen yields only 34,643 BTUs, or 27.6 percent of the energy in a gallon of gasoline. The space shuttle uses hydrogen as a fuel because hydrogen's mass is so low.

- The energy per unit volume is used to determine a fuel's energy density in automobiles. Compressed gaseous hydrogen is even less dense than liquid hydrogen. At 5,000 psi, gaseous hydrogen only has a density of 0.25 pounds per gallon, or 1/24 the density of gasoline. Gasoline and diesel are far superior fuels to hydrogen in this regard.

- The current cost of buying the electricity used in electrolyzing an amount of hydrogen having the energy equal to one gallon-of-gasoline-equivalent (gge) is calculated as follows:

  a. 1 kWh (kilowatt-hour) equals 1,000 J/sec x 3,600 sec = 3.6 million joules/hr.

  b. 237.13 kJ/mole ÷ 3.6 MJ/kWh = 0.06587 kWh/mole.

  c. 1 kilogram (2.2 lbs) of hydrogen is approximately equal to 1 gallon of gasoline in available energy content.

  d. Since 1 mole of hydrogen weighs 2 grams, 1 gallon of gasoline is, therefore, equivalent to 500 moles of hydrogen.

  e. The electric power required to electrolyze the hydrogen energy equivalent to 1 gallon of gasoline is equal to (500 moles) x (0.06667 kWh/mole) = 33.3 kWh, and the approximate cost of the power = (33.3 kWh) (@ 7¢/kWh) = $2.33 per gge. This is a realistic estimate at today's cost of electricity. The following relationships can be readily derived: since 1 kg of hydrogen = 1.04 gallons of gasoline (approx.), then 1 gallon of gasoline is about equal to 961.5 grams of hydrogen; since 1 gallon of liquid hydrogen is about equal to 0.279 gal. of gasoline, then 3.584 gal. of liquid hydrogen about equal to 1 gallon of gasoline.

- In terms of energy use, 1 gigawatt (GWe) of power—the output of most light water nuclear reactors—corresponds to approximately 0.29 million tons/year (Mtons) of hydrogen. One terawatt-year (TW-yr) of energy is equivalent to 0.29 gigatons (Gtons) of hydrogen or 5.13 billion barrels (BB) of oil. The 3.3 TW use of fossil fuels in 2000 would thus correspond to approximately 0.95 Gtons of hydrogen.[1]

Notes

Preface

1. Charlie Osolin, Lawrence Livermore National Laboratory. *Everything Counts: the Multiple Paths to a Carbon-Free Energy Future.* 2005. http://www.llnl.gov/pao/news/news_releases/2005/SF-05-10-01.html

Introduction

1. James L. Williams, Oil Price History and Analysis. 2006. http://www.wtrg.com/prices.htm

2. Robert Bamberger, *Automobile and Light Truck Fuel Economy: The CAFE Standards*, Congressional Research Service, updated September 25, 2002. http://www.policyalmanac.org/environment/archive/ crs_cafe_standards.shtml

3. CNN.com, *Bush adviser: Iraq policy more than "stay the course,"* 2006. http://www.cnn.com/2006/POLITICS/10/23/us.iraq/index.html

4. Energize America; *Energy Security by 2020; Energy Freedom by 2040.* http://www.ea2020.org/drupal/node

5. U.S. Department of Energy, *Hydrogen Posture Plan*, 2004. Page5. http://www1.eere.energy.gov/hydrogenandfuelcells/pdfs/ hydrogen_posture_plan.pdf

5a. John Watson, 1997. U.S. Geological Survey. *What Is Acid Rain?* 1997. http://pubs.usgs.gov/gip/acidrain/2.html

6. Department of Energy Web Site, *FutureGen Project Launched.* 2005
www.fossil.energy.gov/news/techlines/2005/tl_futuregen_signing.html

7. Department of Energy, *President's Hydrogen Fuel Initiative*, 2003
http://www1.eere.energy.gov/hydrogenandfuelcells/
presidents_initiative.html

8. The Coming Global Oil Crisis, *The Hubbert Peak for World Oil*, 2003.
http://www.hubbertpeak.com/summary.htm

9. Jeff Wise, *The Truth About Hydrogen*, Popular Mechanics, 2006. See
"Total Cost" row in Table on page 3.
http://www.popularmechanics.com/technology/industry/
4199381.html?page=3

10. James W Bunger, Peter M. Crawford, Harry R. Johnson, *Is Oil Shale
America's Answer to Peak Oil Challenge?* Oil and Gas Journal. 2004. Page
1. See Map
http://www.fossil.energy.gov/programs/reserves/publications/Pubs-NPR/
40010-373.pdf

11. Julian T. Rubin, *Electrolysis K-12 Experiments & Background Information*, Science Experiment Encyclopedia, 2006.
http://juliantrubin.com/encyclopedia/chemistry/electrolysis.html

12. *Electrochemical Hydrogen Compressor*, National Research Council Canada; 2003;
http://www.nrc-cnrc.gc.ca/highlights/2003/0309psi_e.html

13. *Hydrogen Production, Distribution and Delivery*, Hydrogen Workshop
for Fleet Operators, Slides 3 through 13, 2005.
http://www.hydrogenassociation.org/general/fleet_Module2.pdf

14. *Hydrogen Production, Distribution and Delivery*, Hydrogen Workshop for Fleet Operators, Slides 22 through 30, 2005.
http://www.hydrogenassociation.org/general/fleet_Module2.pdf

15. *Hydrogen Powertrains and Vehicles*, Hydrogen Workshop for Fleet Operators, Slides 2 through 17, 2005.
http://www.hydrogenassociation.org/general/fleet_Module2.pdf

16. *Hydrogen Powertrains and Vehicles*, Hydrogen Workshop for Fleet Operators, Slides 18 through 25, 2005.
http://www.hydrogenassociation.org/general/fleet_Module2.pdf

17. Flynn Research Inc., *Defining the Motors and Generators of Tomorrow*, PPMT Technology, 2006.
http://www.flynnresearch.net/technology/PPMT%20Technology.htm

Chapter 1

1. George S. Stanford, *Integral Fast Reactors: Source of Safe, Abundant, Non-Polluting Power*, The National Center for Public Policy Research, 2001.
http://www.nationalcenter.org/NPA378.html

2. Uranium Information Centre Ltd., Radioactive Waste Management, GPO Box 1649N, Melbourne 3001, Australia, 2002.
http://www.uic.com.au/wast.htm

3. S.M. Stacy, Idaho National Laboratory, *Proving the Principle*, Chapter 24: "The Uranium Trail Fades." Pages 234 to 243, 1999.
http://www.inl.gov/proving-the-principle/chapter_24.pdf

3a. S.M. Stacy, Idaho National Laboratory, *The Uranium Trail Fades*, 1999.
http://www.inl.gov/proving-the-principle/chapter_24.pdf

4. OTA-BP-ENV-126, NTIS order #PB94-173994, GPO stock #052-003-01370-3
Technical Options for the Advanced Liquid Metal Reactor, May 1994. Pages 12 through 16.
http://govinfo.library.unt.edu/ota/Ota_1/DATA/1994/9434.PDF

5. Argonne National Laboratories, *EBR-II cooling tower coming down at Idaho site*, 1998.
http://www.anl.gov/Media_Center/Argonne_News/news98/an981012.html

6. Jeff Wise, *The Truth About Hydrogen*, Popular Mechanics, 2006.
http://www.popularmechanics.com/technology/industry/4199381.html?page=1

7. Ballard Power Systems, *Fuel Cell Technology*, 2005.
http://www.ballard.com/be_informed/fuel_cell_technology/how_the_technology_works

7a. Environmental Defense, *Guess who's funding the global warming doubt shops? 2006.*
http://www.environmentaldefense.org/article.cfm?contentid=4870

8. National Aeronautics and Space Administration, *Recent Warming of Arctic May Affect Worldwide Climate*, 2003.
http://www.nasa.gov/centers/goddard/news/topstory/2003/1023esuice.html

9. The White House, *Fact Sheet: Hydrogen Fuel: a Clean and Secure Energy Future*, 2003.
http://www.whitehouse.gov/news/releases/2003/02/20030206-2.html

10. U.S. Department of Energy, *An Introduction to Argonne National Laboratory's Integral Fast Reactor (IFR) Program.*
http://www.nuc.berkeley.edu/designs/ifr/anlw.html

10a. Charles Till, Center for Reactor Information, *Plentiful Energy and the IFR Story*, 2005
http://www.sustainablenuclear.org/PADs/pad0509till.html

10b. World Nuclear Association, *The Nuclear Fuel Cycle*, 2007
http://www.world-nuclear.org/info/inf03.html

11. U.S. Department of Energy, *Advanced Fuel Cycle Initiative.* 2006.
http://www.ne.doe.gov/pdfFiles/
fy06AfciComparisonReportToCongress.pdf

12. Uranium Information Centre Ltd, GPO Box 1649N, Melbourne 3001, Australia, "The 'Front End' of the Nuclear Fuel Cycle", *Nuclear Electricity,* 7th edition, Chapter 4, 2003.
http://www.uic.com.au/ne4.PDF

13. U.S. Department of Energy, *Global Nuclear Energy Partnership Strategic Plan*, 2007.
http://www.gnep.energy.gov/pdfs/gnepStrategicPlanJanuary2007.pdf

14. Uranium Information Centre Ltd, GPO Box 1649N, Melbourne 3001, Australia, *Fast Neutron Reactors*, 2006
http://www.uic.com.au/nip98.htm

15. The American Nuclear Society; position statement; 2005
Fast Reactor Technology: A Path to Long-Term Energy Sustainability
http://www.ans.org/pi/ps/docs/ps74.pdf

16. Jeff Moskin, *The Integral Fast Reactor*, 2005.Page 2.
http://www.commongroundcommonsense.org/forums/lofiversion/
index.php/t37819.html

17. David Baurac, Argonne National Laboratory, *Passively safe reactors rely on nature to keep them cool*, 1986.
http://www.anl.gov/Media_Center/logos20-1/passive01.htm

18. Jeff Moskin, *The Integral Fast Reactor*, 2005.
http://www.commongroundcommonsense.org/forums/lofiversion/
index.php/t37819.html

Chapter 2

1. STAR-LM Lead-Cooled Closed Fuel Cycle Fast Reactor; 2005;
Search on: Supercritical Steam Cycle for Lead Cooled Nuclear Systems

1a. Robert Collier, *Coaxing oil from huge U.S. shale deposits*, San Francisco Chronicle, 2006.
http://www.sfgate.com/cgi-bin/article.cgi?file=/c/a/2006/09/04/
MNGIEKV0D41.DTL

2. Anton Moisseytsev and James J. Sienicki, *Supercritical CO$_2$Brayton Cycle Control Strategy for Autonomous Liquid Metal-Cooled Reactors*, Argonne National Laboratory, 2004.
http://www.osti.gov/bridge/servlets/purl/840371-mR3VlE/native/
840371.pdf

2a. Anton Moisseytsev and James J. Sienicki, *Supercritical CO2Brayton Cycle Control Strategy for Autonomous Liquid Metal-Cooled Reactors*, Argonne National Laboratory, 2004.
http://www.osti.gov/bridge/servlets/purl/840371-mR3VlE/native/
840371.pdf

3. U.S. Senate Committee, Energy and Natural Resources, Oil Shale and Oil Sands Resources Hearing, Tuesday, April 12, 2005
Search on: TESTIMONY OF STEPHEN MUT

4. U.S. Department of Energy, *How Much Nuclear Waste Is In The United States?*
http://www.ocrwm.doe.gov/ym_repository/about_project/
waste_explained/howmuch.shtml

5. Shell in the US—Mahogany Research Project; Shell Oil Co.; 2006
http://www.shell.com mahogany

6. U.S. Department of the Interior, Bureau of Land Management, *About Oil Shale*, 2005.
http://ostseis.anl.gov/guide/oilshale/index.cfm

7. Bureau of Land Management, *Environmental Assessment; Shell Oil Shale Research, Development, and Demonstration Projects, Rio Blanco County, Colorado*; August 2006
http://www.co.blm.gov/wrra/documents/co1102006117eat.pdf

8. New York State Energy Research and Development Authority, HYDROGEN FACT SHEET, Hydrogen Production—Nuclear, 2006.
http://www.getenergysmart.org/Files/HydrogenEducation/
7HydrogenProductionNuclear.pdf

9. Hydrogen Workshop for Fleet Operators; Shell; 2005;
http://www.hydrogenassociation.org/general/fleet_Module2.pdf

10. Bruce Kinzey, *DOE Hydrogen Program Hydrogen Safety Education and Training for Emergency Responders*, Pacific Northwest National Laboratory (PNNL), 2006.
http://www.hydrogen.energy.gov/pdfs/review06/ed_3_kinzey.pdf

11. Worldwide Hydrogen Fueling Stations
http://www.fuelcells.org/info/charts/h2fuelingstations.pdf

12. U.S. Department of Energy, Energy Information Administration, *Official Energy Statistics from the U.S. Government*, 2005.
http://www.eia.doe.gov/neic/quickfacts/quickoil.html

13. U.S. Department of Energy, Advanced Technologies and Energy Efficiency, 2007.
http://www.fueleconomy.gov/feg/atv.shtml

14. U.S. Environmental Protection Agency, Fuel Cells and Vehicles, 2007.
http://www.epa.gov/fuelcell/basicinfo.htm

15. Ballard Power Systems, *Fuel Cell Technology*, 2005.
http://www.ballard.com/be_informed/fuel_cell_technology/how_the_technology_works

16. U.S. Department of the Interior, Bureau of Land Management (BLM), *Oil Shale and Tar Sands Leasing PEIS Maps*,
http://ostseis.anl.gov/guide/maps/index.cfm

Chapter 3

1. Dr. Sam Vaknin, The Economies of the Middle East, 2003.
http://samvak.tripod.com/brief-middleeast01.html

2. Infoplease, Gross Domestic Product or Expenditure, 1930—2005
http://www.infoplease.com/ipa/A0104575.html

3. U.S. Department of Energy, A Technology Roadmap for Generation IV Nuclear Energy Systems, 2002, Page 85.
Search on:_gif.inel.gov/roadmap/pdfs/gen_iv_roadmap.pdf

4. STAR-LM Lead-Cooled Closed Fuel Cycle Fast Reactor; 2005;
Search on: Supercritical Steam Cycle for Lead Cooled Nuclear Systems

5. Anton Moisseytsev and James J. Sienicki, *Supercritical CO2Brayton Cycle Control Strategy for Autonomous Liquid Metal-Cooled Reactors*, Argonne National Laboratory, 2004.
http://www.osti.gov/bridge/servlets/purl/840371-mR3VlE/native/840371.pdf

6. Jeff Wise, *The Truth About Hydrogen*, Popular Mechanics, 2006. See Table Page 3.
http://www.popularmechanics.com/technology/industry/4199381.html?page=1

7. U.S. Department of Energy, A Technology Roadmap for Generation IV Nuclear Energy Systems, 2002, Page 13.
Search on:_gif.inel.gov/roadmap/pdfs/gen_iv_roadmap.pdf

8. Flynn Research Inc., Defining the Motors and Generators of Tomorrow,
PPMT Technology, 2006.
http://www.flynnresearch.net/technology/PPMT%20Technology.htm

9. Lynn Yarris, A Boost for Hydrogen Fuel Cell Research, 2007.
http://www.lbl.gov/Science-Articles/Archive/MSD-H-fuel-cells.html

10. National Research Council Canada; electrochemical hydrogen compressor; 2003;
http://www.nrc-cnrc.gc.ca/multimedia/picture/environment/nrc-ifci_hydrogen_comp_e.html

10a. Department of Health and Human Services, Atmospheric pollution and the prevalence of asthma: study among schoolchildren of 2 areas in Rio de Janeiro, Brazil, 2004.

http://www.ncbi.nlm.nih.gov/entrez/query.fcgi?cmd=Retrieve&db=
PubMed&list_uids=15237764&dopt=Citation

11. Osman Chughtai and David Shannon, Fossil Fuels, University of
Michigan
http://www.umich.edu/~gs265/society/fossilfuels.htm

Chapter 4

1. U.S. Department of Energy, *How Much Nuclear Waste Is In The United
States?* 2007.
http://www.ocrwm.doe.gov/ym_repository/about_project/
waste_explained/howmuch.shtml

2. Julian Rubin, *Electrolysis K-12 Experiments & Background Information*,
Science Experiment Encyclopedia, 2006.
http://juliantrubin.com/encyclopedia/chemistry/electrolysis.html

3. High-temperature electrolysis of steam; DOE; 2006
http://www.hydrogen.energy.gov/pdfs/progress06/ii_h_6_balachov.pdf

4. National Research Council Canada; electrochemical hydrogen compressor; 2003;
http://www.nrc-cnrc.gc.ca/highlights/2003/0309psi_e.html

5. FuelCellStore.com, Hydrogen Storage, 2006.
http://www.fuelcellstore.com/cgi-bin/fuelweb/view=NavPage/cat=1014

6. Hydrogen Workshop for Fleet Operators; Shell; 2005;
http://www.hydrogenassociation.org/general/fleet_Module2.pdf

7. Parallel Path Magnetic Technology, Patents and Copyright;
www.flynnresearch.net/technology/
PPMT%20technology%20white%20paper.pdf

Chapter 5

1. Department of Energy. *A Technology Roadmap for Generation IV Nuclear Energy Systems*. 2002. Page 16.
Search on: gif.inel.gov/roadmap/pdfs/gen_iv_roadmap.pdf

2. U.S. Department of Energy Hydrogen Posture Plan, 2004. Pages iii and iv.
http://www1.eere.energy.gov/hydrogenandfuelcells/pdfs/
hydrogen_posture_plan.pdf

3. U.S. Department of Energy, A Technology Roadmap for Generation IV Nuclear Energy Systems, 2002. Page 16.
Search on: gif.inel.gov/roadmap/pdfs/gen_iv_roadmap.pdf

4. Finis Southworth, *Very High Temperature Gas Cooled Reactor Systems*, U.S. Department of Energy, Idaho National Engineering & Environmental Laboratory, 2002.
http://gif.inel.gov/roadmap/pdfs/
very_high_temp_gas_cooled_reactor_sys.pdf

5. George S. Stanford, *Integral Fast Reactors: Source of Safe, Abundant, Non-Polluting Power*, The National Center for Public Policy Research, 2001.
http://www.nationalcenter.org/NPA378.html

6. Shell Oil Co, About Mahogany Research Project, 2006.
http://www.shell mahogany

7. U.S. Department of the Interior, Bureau of Land Management, About Oil Shale, 2005.
http://ostseis.anl.gov/guide/oilshale/index.cfm

8. U.S. Department of Energy Hydrogen Program. 2006. Page 1
http://www.hydrogen.energy.gov

9. U.S. Department of Energy. *A National Vision of America's Transition to a Hydrogen Economy—2030 and Beyond*, 2002. Page 11.
http://www1.eere.energy.gov/hydrogenandfuelcells/pdfs/vision_doc.pdf

10. U.S. Department of Energy. Basic Research Needs for a Hydrogen Economy. Second Printing 2004. Page 9
http://www.sc.doe.gov/bes/hydrogen.pdf

11. High-temperature electrolysis of steam; DOE; 2006.
http://www.hydrogen.energy.gov/pdfs/progress06/ii_h_6_balachov.pdf

12. FuelCellStore.com, Hydrogen Storage, 2006.
http://www.fuelcellstore.com/cgi-bin/fuelweb/view=NavPage/cat=1014

13. Alexandra Baker, The Fuel Cell Today, 2005.
Search on: fuelcelltoday.com/FuelCellToday/FCTFiles/FCTArticleFiles/Article_940_GM%20Sequel.pdf

14. National Research Council Canada; electrochemical hydrogen compressor; 2003;
http://www.nrc-cnrc.gc.ca/highlights/2003/0309psi_e.html

15. U.S. Department of Energy, *Hydrogen Posture Plan*, 2004. Page 39.
http://www1.eere.energy.gov/hydrogenandfuelcells/pdfs/hydrogen_posture_plan.pdf

16. Hydrogen Workshop for Fleet Operators, 2005.
http://www.hydrogenassociation.org/general/fleet_Module2.pdf

17. Air Liquide, Press Releases, 2006
http://www.airliquide.com/en/press/press-releases/bayport-hydrogen.html

18. Honda Worldwide, *Home Hydrogen Refueling Technology Advances with the Introduction of Honda's Experimental Home Energy Station*, 2005. http://world.honda.com/news/2005/c051114.html

19. Georgia Tech Research News, *Flying on Hydrogen: Georgia Tech Researchers Use Fuel Cells to Power Unmanned Aerial Vehicle*, 2006. http://gtresearchnews.gatech.edu/newsrelease/fuel-cell-aircraft.htm

20. Otis Port, Hydrogen Cars Are Almost Here, But ..., BusinessWeek Online, 2005.
http://www.businessweek.com/magazine/content/05_04/b3917097_mz018.htm

21. Green Car Congress, *Ford to Begin Production of Hydrogen ICE Shuttles*, 2006.
http://www.greencarcongress.com/2006/06/ford_to_begin_p.html

22. Institute for the Analysis of Global Security, *Fuel Cell Buses to UK and Down Under*, 2004.
http://www.iags.org/n012104t3.htm

23. U.S. Department of Energy's Hydrogen Program Annual Progress Report for Fiscal Year 2005.
http://www.hydrogen.energy.gov/pdfs/progress05/i_introduction.pdf

24. U.S. Department of Energy, *President's Hydrogen Fuel Initiative*, 2003.
http://www1.eere.energy.gov/hydrogenandfuelcells/presidents_initiative.html

25. Ballard Power Systems; Fuel Cell Technology; 2005
http://www.ballard.com/be_informed/fuel_cell_technology/demonstration_programs

26. Iran Daily, Pumping Up Hydrogen's Image, 2004. Page 8.
http://www.irandaily.ir/1383/2139/html/energy.htm

27. David Ivanovich-Houston Chronicle, *Government officials see hydrogen as fuel of future, but cost is high*, Fuel Cell Works, 2004.
http://www.fuelcellsworks.com/Supppage1485.html

28. Tor Hammerstad, Hydro opens world's first hydrogen station on Iceland, 2003.
http://www.hydro.com/en/press_room/news/archive/2003_04/hydrogen_island_en.html

29. Lynn Yarris, *A Boost for Hydrogen Fuel Cell Research*, 2007.
http://www.lbl.gov/Science-Articles/Archive/MSD-H-fuel-cells.html

30. Flynn Research Inc., Defining the Motors and Generators of Tomorrow, PPMT Technology, 2006.
http://www.flynnresearch.net/technology/PPMT%20Technology.htm

31. TIAX LLC, Cambridge, MA 02140-2390, *PEM Fuel Cell Cost Status*, 2005.
http://www.fuelcellseminar.com/pdf/2005/Thursday-Nov17/Carlson_Eric_392.PDF

31a. EVWORLD, *Nanonickel to Replace Platinum as Fuel Cell Catalyst*, 2005
http://www.evworld.com/news.cfm?newsid=8351

32. Greg Schneider, *Priming the Public For Hydrogen Fuel, Benning Road Station Is First of Its Kind in U.S.*, Washington Post, 2004.
http://www.washingtonpost.com/wp-dyn/articles/A38168-2004Nov9.html

Appendix

1. U.S. Department of Energy. Basic Research Needs for a Hydrogen Economy. Second Printing 2004. Page 9, Footnote 1. http://www.sc.doe.gov/bes/hydrogen.pdf

978-0-595-43519-7
0-595-43519-X